"十三五"生态环境保护规划

中国环境出版社·北京

图书在版编目（CIP）数据

"十三五"生态环境保护规划 / 环境保护部环境规划院编． -- 北京：中国环境出版社，2017.5
　ISBN 978-7-5111-3181-2

　Ⅰ．①十… Ⅱ．①环… Ⅲ．①生态环境－环境规划－中国－2016-2020 Ⅳ．①X321.2

中国版本图书馆CIP数据核字（2017）第110339号

出 版 人	王新程
责任编辑	陶克菲
策划编辑	刘亚庚
责任校对	尹　芳
装帧设计	宋　瑞

出版发行　中国环境出版社
　　　　　（100062　北京市东城区广渠门内大街16号）
　　　　　网　　址：http://www.cesp.com.cn
　　　　　电子邮箱：bjgl@cesp.com.cn
　　　　　联系电话：010-67112765（编辑管理部）
　　　　　发行热线：010-67125803，010-67113405（传真）
　　　　　印装质量热线：010-67113404

印　　刷	北京中科印刷有限公司
经　　销	各地新华书店
版　　次	2017年5月第1版
印　　次	2017年5月第1次印刷
开　　本	787×960　1/16
印　　张	12.75
字　　数	110千字
定　　价	32.00元

【版权所有。未经许可，请勿翻印、转载，违者必究】
如有缺页、破损、倒装等印装质量问题，请寄回本社更换

目 录

第一章　全国生态环境保护形势..2
　　第一节　生态环境保护取得积极进展..2
　　第二节　生态环境是全面建成　小康社会的突出短板..........................5
　　第三节　生态环境保护面临机遇与挑战..6

第二章　指导思想、基本原则与主要目标......................................8
　　第一节　指导思想..8
　　第二节　基本原则..9
　　第三节　主要目标...10

第三章　强化源头防控，夯实绿色发展基础...................................13
　　第一节　强化生态空间管控...13
　　第二节　推进供给侧结构性改革...15
　　第三节　强化绿色科技创新引领...17
　　第四节　推动区域绿色协调发展...20

第四章　深化质量管理，大力实施三大行动计划...............................23
　　第一节　分区施策改善大气环境质量..23
　　第二节　精准发力提升水环境质量..26
　　第三节　分类防治土壤环境污染..32

第五章 实施专项治理，全面推进达标排放与污染减排 ………… 36
第一节 实施工业污染源全面达标排放计划 ………… 36
第二节 深入推进重点污染物减排 ………… 38
第三节 加强基础设施建设 ………… 44
第四节 加快农业农村环境综合治理 ………… 46

第六章 实行全程管控，有效防范和降低环境风险 ………… 49
第一节 完善风险防控和应急响应体系 ………… 49
第二节 加大重金属污染防治力度 ………… 51
第三节 提高危险废物处置水平 ………… 53
第四节 夯实化学品风险防控基础 ………… 54
第五节 加强核与辐射安全管理 ………… 54

第七章 加大保护力度，强化生态修复 ………… 56
第一节 维护国家生态安全 ………… 56
第二节 管护重点生态区域 ………… 57
第三节 保护重要生态系统 ………… 58
第四节 提升生态系统功能 ………… 59
第五节 修复生态退化地区 ………… 60
第六节 扩大生态产品供给 ………… 61
第七节 保护生物多样性 ………… 63

第八章 加快制度创新，积极推进治理体系和治理能力现代化 ………… 65
第一节 健全法治体系 ………… 65
第二节 完善市场机制 ………… 66
第三节 落实地方责任 ………… 68
第四节 加强企业监管 ………… 70
第五节 实施全民行动 ………… 71

第六节　提升治理能力..72

第九章　实施一批国家生态环境保护重大工程..................**76**

第十章　健全规划实施保障措施..................**84**
　　第一节　明确任务分工..84
　　第二节　加大投入力度..84
　　第三节　加强国际合作..85
　　第四节　推进试点示范..86
　　第五节　严格评估考核..86

Chapter 1 An Overview of Environmental Protection in China...........91
　　Section 1 Positive progress..91
　　Section 2 Ecological environment remaining a weak link in achieving a moderately prosperous society in all aspects.............94
　　Section 3 Opportunities and challenges for the protection of ecological environment..96

Chapter 2 Guidelines, Basic Principles and Main Objectives................98
　　Section 1 Guidelines..98
　　Section 2 Basic principles..98
　　Section 3 Main objectives..100

Chapter 3 Prevent and Control Pollution at Source to Lay Groundwork for Green Development..104
　　Section 1 Strengthen management and control of ecological space 104
　　Section 2 Advance supply-side structural reform....................106

Section 3 Drive growth through green technology innovation...109

Section 4 Promote green and coordinated regional development 113

Chapter 4 Promote Quality-oriented Management in Implementing the Three Action Plans on Prevention and Control of Air, Water and Soil Pollution...116

Section 1 Improve ambient air quality in line with regional conditions..116

Section 2 Improve water quality with targeted measures............120

Section 3 Tackle soil contamination by category.......................128

Chapter 5 Promote Up-to-Standard Discharge and Emissions Reduction While Targeting Special Sectors and Areas..132

Section 1 Implement plans on up-to-standard discharge from all industrial pollution sources...132

Section 2 Further reduce major pollutants discharge..................134

Section 3 Advance infrastructure construction...........................141

Section 4 Tackle agricultural pollution and improve rural environment in a comprehensive manner.................................145

Chapter 6 Integrate Pollution Prevention and Control into the Whole Process to Effectively Prevent and Mitigate Environmental Risks.......147

Section 1 Improve the systems of risk prevention and control and emergency response..147

Section 2 Step up prevention and control of heavy metal pollution 149

Section 3 Enhance hazardous waste disposal............................152

Section 4 Build up the capacity to manage chemical risks.........153

Section 5 Strengthen management over nuclear and radiation safety..154

Chapter 7 Enhance Ecological Protection and Restoration..................156

Section 1 Safeguard national ecological security......................156

Section 2 Manage and protect key ecological regions..............157

Section 3 Protect important ecosystems....................................158

Section 4 Enhance ecosystem functions....................................160

Section 5 Restore ecologically degraded areas.........................161

Section 6 Expand ecological products supply..........................162

Section 7 Conserve biodiversity..164

Chapter 8 Modernize Governance System and Enhance its Capacity by Accelerating Institutional Innovation..167

Section 1 Put in place a sound legal framework.......................167

Section 2 Improve market mechanism......................................168

Section 3 Ensure localities fulfill their due responsibilities.......170

Section 4 Strengthen regulation of enterprises..........................173

Section 5 Mobilize public support...175

Section 6 Build up governance capacity...................................176

Chapter 9 Implement National Projects on Protecting the Ecological Environment...180

Chapter 10 Introduce Supporting Measures to Facilitate Implementation..191

Section 1 Clarify responsibilities and tasks...............................191

Section 2 Increase inputs...191
Section 3 Expand international cooperation................................192
Section 4 Push ahead pilot and demonstration projects..............193
Section 5 Examine and evaluate progress and performance with strict standards..194

国务院关于印发"十三五"生态环境保护规划的通知

国发〔2016〕65号

各省、自治区、直辖市人民政府,国务院各部委、各直属机构:

现将《"十三五"生态环境保护规划》印发给你们,请认真贯彻实施。

国务院
2016年11月24日

第一章
全国生态环境保护形势

党中央、国务院高度重视生态环境保护工作。"十二五"以来，坚决向污染宣战，全力推进大气、水、土壤污染防治，持续加大生态环境保护力度，生态环境质量有所改善，完成了"十二五"规划确定的主要目标和任务。"十三五"期间，经济社会发展不平衡、不协调、不可持续的问题仍然突出，多阶段、多领域、多类型生态环境问题交织，生态环境与人民群众需求和期待差距较大，提高环境质量，加强生态环境综合治理，加快补齐生态环境短板，是当前核心任务。

第一节 生态环境保护取得积极进展

生态文明建设上升为国家战略。党中央、国务院高度重视生态文明建设。习近平总书记多次强调，"绿水青山就是金山银山"，"要坚持节约资源和保护环境的基本国策"，"像保护眼睛一样保护生态环境，像对待生命一样对待生态环境"。李克强总理多次指出，要加大环境综合治理力度，提高生态文明水平，促进绿色发展，下决心走出一条经济发展与环境改善双赢之路。党的十八大以来，党中央、国务院把生态文明建设摆在更加重要的战略位置，纳入"五位一体"总体布局，作出一系列重大决策部署，出台《生态文明体制改革总体方案》，实施大气、水、土壤污染防治行动计划。把发展观、执政观、自然观内在统一起来，融入到执政理念、发展理念中，生态文明建设的认识高度、实践深度、推进力度前所未有。

生态环境质量有所改善。2015年，全国338个地级及以上城市细颗粒物（$PM_{2.5}$）年均浓度为50微克/立方米，首批开展监测的74个城市细颗粒物年均浓度比2013年下降23.6%，京津冀、长三角、珠三角分别下降27.4%、20.9%、27.7%，酸雨区占国土面积比例由历史高峰值的30%左右降至7.6%，大气污染防治初见成效。全国1940个地表水国控断面Ⅰ—Ⅲ类比例提高至66%，劣Ⅴ类比例下降至9.7%，大江大河干流水质明显改善。全国森林覆盖率提高至21.66%，森林蓄积量达到151.4亿立方米，草原综合植被盖度54%。建成自然保护区2740个，占陆地国土面积14.8%，超过90%的陆地自然生态系统类型、89%的国家重点保护野生动植物种类以及大多数重要自然遗迹在自然保护区内得到保护，大熊猫、东北虎、朱鹮、藏羚羊、扬子鳄等部分珍稀濒危物种野外种群数量稳中有升。荒漠化和沙化状况连续三个监测周期实现面积"双缩减"。

治污减排目标任务超额完成。到2015年，全国脱硫、脱硝机组容量占煤电总装机容量比例分别提高到99%、92%，完成煤电机组超低排放改造1.6亿千瓦。全国城市污水处理率提高到92%，城市建成区生活垃圾无害化处理率达到94.1%。7.2万个村庄实施环境综合整治，1.2亿多农村人口直接受益。6.1万家规模化养殖场（小区）建成废弃物处理和资源化利用设施。"十二五"

期间，全国化学需氧量和氨氮、二氧化硫、氮氧化物排放总量分别累计下降 12.9%、13%、18%、18.6%。

生态保护与建设取得成效。天然林资源保护、退耕还林还草、退牧还草、防护林体系建设、河湖与湿地保护修复、防沙治沙、水土保持、石漠化治理、野生动植物保护及自然保护区建设等一批重大生态保护与修复工程稳步实施。重点国有林区天然林全部停止商业性采伐。全国受保护的湿地面积增加 525.94 万公顷，自然湿地保护率提高到 46.8%。沙化土地治理 10 万平方公里、水土流失治理 26.6 万平方公里。完成全国生态环境十年变化（2000—2010 年）调查评估，发布《中国生物多样性红色名录》。建立各级森林公园、湿地公园、沙漠公园 4300 多个。16 个省（区、市）开展生态省建设，1000 多个市（县、区）开展生态市（县、区）建设，114 个市（县、区）获得国家生态建设示范区命名。国有林场改革方案及国有林区改革指导意见印发实施，6 个省完成国有林场改革试点任务。

环境风险防控稳步推进。到 2015 年，50 个危险废物、273 个医疗废物集中处置设施基本建成，历史遗留的 670 万吨铬渣全部处置完毕，铅、汞、镉、铬、砷五种重金属污染物排放量比 2007 年下降 27.7%，涉重金属突发环境事件数量大幅减少。科学应对天津港"8·12"特别重大火灾爆炸等事故环境影响。核设施安全水平持续提高，核技术利用管理日趋规范，辐射环境质量保持良好。

生态环境法治建设不断完善。环境保护法、大气污染防治法、放射性废物安全管理条例、环境空气质量标准等完成制修订，生态环境损害责任追究办法等文件陆续出台，生态保护补偿机制进一步健全。深入开展环境保护法实施年活动和环境保护综合督察。

全社会生态环境法治观念和意识不断加强。

第二节 生态环境是全面建成小康社会的突出短板

污染物排放量大面广，环境污染重。我国化学需氧量、二氧化硫等主要污染物排放量仍然处于2000万吨左右的高位，环境承载能力超过或接近上限。78.4%的城市空气质量未达标，公众反映强烈的重度及以上污染天数比例占3.2%，部分地区冬季空气重污染频发高发。饮用水水源安全保障水平亟需提升，排污布局与水环境承载能力不匹配，城市建成区黑臭水体大量存在，湖库富营养化问题依然突出，部分流域水体污染依然较重。全国土壤点位超标率16.1%，耕地土壤点位超标率19.4%，工矿废弃地土壤污染问题突出。城乡环境公共服务差距大，治理和改善任务艰巨。

山水林田湖缺乏统筹保护，生态损害大。中度以上生态脆弱区域占全国陆地国土面积的55%，荒漠化和石漠化土地占国土面积的近20%。森林系统低质化、森林结构纯林化、生态功能低效化、自然景观人工化趋势加剧，每年违法违规侵占林地约200万亩，全国森林单位面积蓄积量只有全球平均水平的78%。全国草原生态总体恶化局面尚未根本扭转，中度和重度退化草原面积仍占1/3以上，已恢复的草原生态系统较为脆弱。全国湿地面积近年来每年减少约510万亩，900多种脊椎动物、3700多种高等植物生存受到威胁。资源过度开发利用导致生态破坏问题突出，生态空间不断被蚕食侵占，一些地区生态资源破坏严重，系统保护难度加大。

产业结构和布局不合理，生态环境风险高。我国是化学品生产和消费大国，有毒有害污染物种类不断增加，区域性、结构性、布局性环境风险日益凸显。环境风险企业数量庞大、近水靠城，危险化学品安全事故导致的环境污染事件频发。突发环境事件呈现原因复杂、污染物质多样、影响地域敏感、影响范围扩大的趋势。过去十年年均发生森林火灾 7600 多起，森林病虫害发生面积 1.75 亿亩以上。近年来，年均截获有害生物达 100 万批次，动植物传染及检疫性有害生物从国境口岸传入风险高。

第三节 生态环境保护面临机遇与挑战

"十三五"期间，生态环境保护面临重要的战略机遇。全面深化改革与全面依法治国深入推进，创新发展和绿色发展深入实施，生态文明建设体制机制逐步健全，为环境保护释放政策红利、法治红利和技术红利。经济转型升级、供给侧结构性改革加快化解重污染过剩产能、增加生态产品供给，污染物新增排放压力趋缓。公众生态环境保护意识日益增强，全社会保护生态环境的合力逐步形成。

同时，我国工业化、城镇化、农业现代化的任务尚未完成，生态环境保护仍面临巨大压力。伴随着经济下行压力加大，发展与保护的矛盾更加突出，一些地方环保投入减弱，进一步推进环境治理和质量改善任务艰巨。区域生态环境分化趋势显现，污染点状分布转向面上扩张，部分地区生态系统稳定性和服务功能下降，统筹协调保护难度大。我国积极应对全球气候变化，推进"一带一路"建设，国际社会尤其是发达国家要求我国承担更多环境责任，深度参与全球环境治理挑战大。

"十三五"期间，生态环境保护机遇与挑战并存，既是负重前行、大有作为的关键期，也是实现质量改善的攻坚期、窗口期。要充分利用新机遇新条件，妥善应对各种风险和挑战，坚定推进生态环境保护，提高生态环境质量。

第二章
指导思想、基本原则与主要目标

第一节 指导思想

全面贯彻党的十八大和十八届三中、四中、五中、六中全会精神，以邓小平理论、"三个代表"重要思想、科学发展观为指导，深入贯彻习近平总书记系列重要讲话精神和治国理政新理念新思想新战略，统筹推进"五位一体"总体布局和协调推进"四个全面"战略布局，牢固树立和贯彻落实创新、协调、绿色、开放、共享的发展理念，按照党中央、国务院决策部署，以提高环境质量为核心，实施最严格的环境保护制度，打好大气、水、土壤污染防治三大战役，加强生态保护与修复，严密防控生态环境风险，加快推进生态环境领域国家治理体系和治理能力现代化，不断提高生态环境管理系统化、科学化、法治化、精细化、信息化水平，为人民提供更多优质生态产品，为实现"两个一百年"奋斗目标和中华民族伟大复兴的中国梦作出贡献。

第二节 基本原则

坚持绿色发展、标本兼治。绿色富国、绿色惠民，处理好发展和保护的关系，协同推进新型工业化、城镇化、信息化、农业现代化与绿色化。坚持立足当前与着眼长远相结合，加强生态环境保护与稳增长、调结构、惠民生、防风险相结合，强化源头防控，推进供给侧结构性改革，优化空间布局，推动形成绿色生产和绿色生活方式，从源头预防生态破坏和环境污染，加大生态环境治理力度，促进人与自然和谐发展。

坚持质量核心、系统施治。以解决生态环境突出问题为导向，分区域、分流域、分阶段明确生态环境质量改善目标任务。统筹运用结构优化、污染治理、污染减排、达标排放、生态保护等多种手段，实施一批重大工程，开展多污染物协同防治，系统推进生态修复与环境治理，确保生态环境质量稳步提升，提高优质生态产品供给能力。

坚持空间管控、分类防治。生态优先，统筹生产、生活、生态空间管理，划定并严守生态保护红线，维护国家生态安全。建立系统完整、责权清晰、监管有效的管理格局，实施差异化管理，分区分类管控，分级分项施策，提升精细化管理水平。

坚持改革创新、强化法治。以改革创新推进生态环境保护，转变环境治理理念和方式，改革生态环境治理基础制度，建立覆盖所有固定污染源的企业排放许可制，实行省以下环保机构监测监察执法垂直管理制度，加快形成系统完整的生态文明制度体系。加强环境立法、环境司法、环境执法，从硬从严，重拳出击，促进全社会遵纪守法。依靠法律和制度加强生态环境保护，实现源头严防、过程严管、后果严惩。

坚持履职尽责、社会共治。建立严格的生态环境保护责任制度，合理划分中央和地方环境保护事权和支出责任，落实生态环境保护"党政同责"、"一岗双责"。落实企业环境治理主体责任，动员全社会积极参与生态环境保护，激励与约束并举，政府与市场"两手发力"，形成政府、企业、公众共治的环境治理体系。

第三节　主要目标

到 2020 年，生态环境质量总体改善。生产和生活方式绿色、低碳水平上升，主要污染物排放总量大幅减少，环境风险得到有效控制，生物多样性下降势头得到基本控制，生态系统稳定性明显增强，生态安全屏障基本形成，生态环境领域国家治理体系和治理能力现代化取得重大进展，生态文明建设水平与全面建成小康社会目标相适应。

专栏1 "十三五"生态环境保护主要指标

指标		2015年	2020年	〔累计〕[1]	属性
生态环境质量					
1.空气质量	地级及以上城市[2]空气质量优良天数比率（%）	76.7	>80	—	约束性
	细颗粒物未达标地级及以上城市浓度下降（%）	—	—	〔18〕	约束性
	地级及以上城市重度及以上污染天数比例下降（%）	—	—	〔25〕	预期性
2.水环境质量	地表水质量[3]达到或好于Ⅲ类水体比例（%）	66	>70	—	约束性
	地表水质量劣Ⅴ类水体比例（%）	9.7	<5	—	约束性
	重要江河湖泊水功能区水质达标率（%）	70.8	>80	—	预期性
	地下水质量极差比例（%）	15.7[4]	15左右	—	预期性
	近岸海域水质优良（一、二类）比例（%）	70.5	70左右	—	预期性
3.土壤环境质量	受污染耕地安全利用率（%）	70.6	90左右	—	约束性
	污染地块安全利用率（%）	—	90以上	—	约束性
4.生态状况	森林覆盖率（%）	21.66	23.04	〔1.38〕	约束性
	森林蓄积量（亿立方米）	151	165	〔14〕	约束性
	湿地保有量（亿亩）	—	≥8	—	预期性
	草原综合植被盖度（%）	54	56	—	预期性
	重点生态功能区所属县域生态环境状况指数	60.4	>60.4	—	预期性

指　　标		2015年	2020年	〔累计〕[1]	属性
污染物排放总量					
5.主要污染物排放总量减少（%）	化学需氧量	—	—	〔10〕	约束性
	氨氮	—	—	〔10〕	
	二氧化硫	—	—	〔15〕	
	氮氧化物	—	—	〔15〕	
6.区域性污染物排放总量减少（%）	重点地区重点行业挥发性有机物[5]	—	—	〔10〕	预期性
	重点地区总氮[6]	—	—	〔10〕	预期性
	重点地区总磷[7]	—	—	〔10〕	
生态保护修复					
7.国家重点保护野生动植物保护率（%）		—	≥95	—	预期性
8.全国自然岸线保有率（%）		—	≥35	—	预期性
9.新增沙化土地治理面积（万平方公里）		—	—	〔10〕	预期性
10.新增水土流失治理面积（万平方公里）		—	—	〔27〕	预期性

注：1.〔　〕内为五年累计数。
　　2.空气质量评价覆盖全国338个城市（含地、州、盟所在地及部分省辖县级市，不含三沙和儋州）。
　　3.水环境质量评价覆盖全国地表水国控断面，断面数量由"十二五"期间的972个增加到1940个。
　　4.为2013年数据。
　　5.在重点地区、重点行业推进挥发性有机物总量控制，全国排放总量下降10%以上。
　　6.对沿海56个城市及29个富营养化湖库实施总氮总量控制。
　　7.总磷超标的控制单元以及上游相关地区实施总磷总量控制。

第三章
强化源头防控，夯实绿色发展基础

绿色发展是从源头破解我国资源环境约束瓶颈、提高发展质量的关键。要创新调控方式，强化源头管理，以生态空间管控引导构建绿色发展格局，以生态环境保护推进供给侧结构性改革，以绿色科技创新引领生态环境治理，促进重点区域绿色、协调发展，加快形成节约资源和保护环境的空间布局、产业结构和生产生活方式，从源头保护生态环境。

第一节 强化生态空间管控

全面落实主体功能区规划。强化主体功能区在国土空间开发保护中的基础作用，推动形成主体功能区布局。依据不同区域主体功能定位，制定差异化的生态环境目标、治理保护措施和考核评价要求。禁止开发区域实施强制性生态环境保护，严格控制人为因素对自然生态和自然文化遗产原真性、完整性的干扰，严禁不符合主体功能定位的各类开发活动，引导人口逐步有序转移。

限制开发的重点生态功能区开发强度得到有效控制，形成环境友好型的产业结构，保持并提高生态产品供给能力，增强生态系统服务功能。限制开发的农产品主产区着力保护耕地土壤环境，确保农产品供给和质量安全。重点开发区域加强环境管理与治理，大幅降低污染物排放强度，减少工业化、城镇化对生态环境的影响，改善人居环境，努力提高环境质量。优化开发区域引导城市集约紧凑、绿色低碳发展，扩大绿色生态空间，优化生态系统格局。实施海洋主体功能区规划，优化海洋资源开发格局。

划定并严守生态保护红线。2017年底前，京津冀区域、长江经济带沿线各省（市）划定生态保护红线；2018年底前，其他省（区、市）划定生态保护红线；2020年底前，全面完成全国生态保护红线划定、勘界定标，基本建立生态保护红线制度。制定生态保护红线管控措施，建立健全生态保护补偿机制，定期发布生态保护红线保护状况信息。建立监控体系与评价考核制度，对各省（区、市）生态保护红线保护成效进行评价考核。全面保障国家生态安全，保护和提升森林、草原、河流、湖泊、湿地、海洋等生态系统功能，提高优质生态产品供给能力。

推动"多规合一"。以主体功能区规划为基础，规范完善生态环境空间管控、生态环境承载力调控、环境质量底线控制、战略环评与规划环评刚性约束等环境引导和管控要求，制定落实生态保护红线、环境质量底线、资源利用上线和环境准入负面清单的技术规范，强化"多规合一"的生态环境支持。以市县级行政区为单元，建立由空间规划、用途管制、差异化绩效考核等构成的空间治理体系。积极推动建立国家空间规划体系，统筹各类空间规划，推进"多规合一"。研究制定生态环境保护促进"多规合一"的指导意见。自2018年起，启动省域、区域、城市群生态环境保护空间规划研究。

第二节 推进供给侧结构性改革

强化环境硬约束推动淘汰落后和过剩产能。建立重污染产能退出和过剩产能化解机制，对长期超标排放的企业、无治理能力且无治理意愿的企业、达标无望的企业，依法予以关闭淘汰。修订完善环境保护综合名录，推动淘汰高污染、高环境风险的工艺、设备与产品。鼓励各地制定范围更宽、标准更高的落后产能淘汰政策，京津冀地区要加大对不能实现达标排放的钢铁等过剩产能淘汰力度。依据区域资源环境承载能力，确定各地区造纸、制革、印染、焦化、炼硫、炼砷、炼油、电镀、农药等行业规模限值。实行新（改、扩）建项目重点污染物排放等量或减量置换。调整优化产业结构，煤炭、钢铁、水泥、平板玻璃等产能过剩行业实行产能等量或减量置换。

严格环保能耗要求促进企业加快升级改造。实施能耗总量和强度"双控"行动，全面推进工业、建筑、交通运输、公共机构等重点领域节能。严格新建项目节能评估审查，加强工业节能监察，强化全过程节能监管。钢铁、有色金属、化工、建材、轻工、纺织等传统制造业全面实施电机、变压器等能效提升、清洁生产、节水治污、循环利用等专项技术改造，实施系统能效提升、燃煤锅炉节能环保综合提升、绿色照明、余热暖民等节能重点工程。支持企业增强绿色精益制造能力，推动工业园区和企业应用分布式能源。

促进绿色制造和绿色产品生产供给。从设计、原料、生产、采购、物流、回收等全流程强化产品全生命周期绿色管理。支持企业推行绿色设计，开发绿色产品，完善绿色包装标准体系，推动包装减量化、无害化和材料回收利用。建设绿色工厂，发展绿色工业园区，打造绿色供应链，开展绿色评价和绿色制造工艺推广行动，全面推进绿色制造体系建设。增强绿色供给能力，整合环保、节能、节水、循环、低碳、再生、有机等产品认证，建立统一的绿色产品标准、认证、标识体系。发展生态农业和有机农业，加快有机食品基地建设和产业发展，增加有机产品供给。到2020年，创建百家绿色设计示范企业、百家绿色示范园区、千家绿色示范工厂，绿色制造体系基本建立。

推动循环发展。实施循环发展引领计划，推进城市低值废弃物集中处置，开展资源循环利用示范基地和生态工业园区建设，建设一批循环经济领域国家新型工业化产业示范基地和循环经济示范市县。实施高端再制造、智能再制造和在役再制造示范工程。深化工业固体废物综合利用基地建设试点，建设产业固体废物综合利用和资源再生利用示范工程。依托国家"城市矿产"示范基地，培育一批回收和综合利用骨干企业、再生资源利用产业基地和园区。健全再生资源回收利用网络，规范完善废钢铁、废旧轮胎、废旧纺织品与服装、废塑料、废旧动力电池等综合利用行业管理。尝试建立逆向回收渠道，推广"互联网＋回收"、智能回收等新型回收方式，实行生产者责任延伸制度。到2020年，全国工业固体废物综合利用率提高到73%。实现化肥农药零增长，实施循环农业示范工程，推进秸秆高值化和产业化利用。到2020年，秸秆综合利用率达到85%，国家现代农业示范区和粮食主产县基本实现农业资源循环利用。

推进节能环保产业发展。推动低碳循环、治污减排、监测监控等核心环保技术工艺、成套产品、装备设备、材料药剂研发与产业化，尽快形成一批具有竞争力的主导技术和产品。鼓励发展节能环保技术咨询、系统设计、设备制造、工程施工、运营管理等专业化服务。大力发展环境服务业，推进形成合同能源管理、合同节水管理、第三方监测、环境污染第三方治理及环境保护政府和社会资本合作等服务市场，开展小城镇、园区环境综合治理托管服务试点。规范环境绩效合同管理，逐步建立环境服务绩效评价考核机制。发布政府采购环境服务清单。鼓励社会资本投资环保企业，培育一批具有国际竞争力的大型节能环保企业与环保品牌。鼓励生态环保领域大众创业、万众创新。充分发挥环保行业组织、科技社团在环保科技创新、成果转化和产业化过程中的作用。完善行业监管制度，开展环保产业常规调查统计工作，建立环境服务企业诚信档案，发布环境服务业发展报告。

第三节　强化绿色科技创新引领

推进绿色化与创新驱动深度融合。把绿色化作为国家实施创新驱动发展战略、经济转型发展的重要基点，推进绿色化与各领域新兴技术深度融合发展。发展智能绿色制造技术，推动制造业向价值链高端攀升。发展生态绿色、高效安全的现代农业技术，深入开展节水农业、循环农业、有机农业、现代林业和生物肥料等技术研发，促进农业提质增效和可持续发展。发展安全、清洁、高效的现代能源技术，推动能源生产和消费革命。发展资源节约循环利用的关键技术，建立城镇生活垃圾资源化利用、再生资源回收利用、工业固体废物综合利用等技术体系。重点针对大气、水、土壤等问题，形成源头预防、末端治理和生态环境修复的成套技术。

加强生态环保科技创新体系建设。瞄准世界生态环境科技发展前沿，立足我国生态环境保护的战略要求，突出自主创新、综合集成创新，加快构建层次清晰、分工明确、运行高效、支撑有力的国家生态环保科技创新体系。重点建立以科学研究为先导的生态环保科技创新理论体系，以应用示范为支撑的生态环保技术研发体系，以人体健康为目标的环境基准和环境标准体系，以提升竞争力为核心的环保产业培育体系，以服务保障为基础的环保科技管理体系。实施环境科研领军人才工程，加强环保专业技术领军人才和青年拔尖人才培养，重点建设一批创新人才培养基地，打造一批高水平创新团队。支持相关院校开展环保基础科学和应用科学研究。建立健全环保职业荣誉制度。

建设生态环保科技创新平台。统筹科技资源，深化生态环保科技体制改革。加强重点实验室、工程技术中心、科学观测研究站、环保智库等科技创新平台建设，加强技术研发推广，提高管理科学化水平。积极引导企业与科研机构加强合作，强化企业创新主体作用，推动环保技术研发、科技成果转移转化和推广应用。推动建立环保装备与服务需求信息平台、技术创新转化交易平台。依托有条件的科技产业园区，集中打造环保科技创新试验区、环保高新技术产业区、环保综合治理技术服务区、国际环保技术合作区、环保高水平人才培养教育区，建立一批国家级环保高新技术产业开发区。

实施重点生态环保科技专项。继续实施水体污染控制与治理国家科技重大专项，实施大气污染成因与控制技术研究、典型脆弱生态修复与保护研究、煤炭清洁高效利用和新型节能技术研发、农业面源和重金属污染农田综合防治与修复技术研究、海洋环境安全保障等重点研发计划专项。在京津冀地区、长江经济带、"一带一路"沿线省（区、市）等重点区域开展环境污染防治和生态修复技术应用试点示范，提出生态环境治理系统性技术解决方案。打造京津冀等区域环境质量提升协同创新共同体，实施区域环境质量提升创新科技工程。创新青藏高原等生态屏障带保护修复技术方法与治理模式，研发生态环境监测预警、生态修复、生物多样性保护、生态保护红线评估管理、生态廊道构建等关键技术，建立一批生态保护与修复科技示范区。支持生态、土壤、大气、温室气体等环境监测预警网络系统及关键技术装备研发，支持生态环境突发事故监测预警及应急处置技术、遥感监测技术、数据分析与服务产品、高端环境监测仪器等研发。开展重点行业危险废物污染特性与环境效应、危险废物溯源及快速识别、全过程风险防控、信息化管理技术等领域研究，加快建立危险废物技术规范体系。建立化学品环境与健康风险评估方法、程序和技术规范体系。加强生态环境管理决策支撑科学研究，开展多污染物协同控制、生态环境系统模拟、污染源解析、生态环境保护规划、生态环境损害评估、网格化管理、绿色国内生产总值核算等技术方法研究应用。

完善环境标准和技术政策体系。研究制定环境基准，修订土壤环境质量标准，完善挥发性有机物排放标准体系，严格执行污染物排放标准。加快机动车和非道路移动源污染物排放标准、燃油产品质量标准的制修订和实施。发布实施船舶发动机排气污染物排放限值及测量方法（中国第一、二阶段）、轻型汽车和重型汽车污染物排放限值及测量方法（中国第六阶段）、摩托车和轻便摩托车污染物排放限值及测量方法（中国第四阶段）、畜禽养殖污染物排放标准。修订在用机动车排放标准，力争实施非道路移动机械国Ⅳ排放标准。完善环境保护技术政策，建立生态保护红线监管技术规范。健全钢铁、水泥、化工等重点行业清洁生产评价指标体系。加快制定完善电力、冶金、有色金属等重点行业以及城乡垃圾处理、机动车船和非道路移动机械污染防治、农业面源污染防治等重点领域技术政策。建立危险废物利用处置无害化管理标准和技术体系。

第四节 推动区域绿色协调发展

促进四大区域绿色协调发展。西部地区要坚持生态优先，强化生态环境保护，提升生态安全屏障功能，建设生态产品供给区，合理开发石油、煤炭、天然气等战略性资源和生态旅游、农畜产品等特色资源。东北地区要加强大小兴安岭、长白山等森林生态系统保护和北方防沙带建设，强化东北平原湿地和农用地土壤环境保护，推动老工业基地振兴。中部地区要以资源环境承载能力为基础，有序承接产业转移，推进鄱阳湖、洞庭湖生态经济区和汉江、淮河生态经济带建设，研究建设一批流域沿岸及交通通道沿线的生态走廊，加强水环境保护和治理。东部地区要扩大生态空间，提高环境资源利用效率，加快推动产业升级，在生态环境质量改善等方面走在前列。

推进"一带一路"绿色化建设。加强中俄、中哈以及中国—东盟、上海合作组织等现有多双边合作机制，积极开展澜沧江—湄公河环境合作，开展全方位、多渠道的对话交流活动，加强与沿线国家环境官员、学者、青年的交流和合作，开展生态环保公益活动，实施绿色丝路使者计划，分享中国生态文明、绿色发展理念与实践经验。建立健全绿色投资与绿色贸易管理制度体系，落实对外投资合作环境保护指南。开展环保产业技术合作园区及示范基地建设，推动环保产业走出去。树立中国铁路、电力、汽车、通信、新能源、钢铁等优质产能绿色品牌。推进"一带一路"沿线省（区、市）产业结构升级与创新升级，推动绿色产业链延伸；开展重点战略和关键项目环境评估，提高生态环境风险防范与应对能力。编制实施国内"一带一路"沿线区域生态环保规划。

推动京津冀地区协同保护。以资源环境承载能力为基础，优化经济发展和生态环境功能布局，扩大环境容量与生态空间。加快推动天津传统制造业绿色化改造。促进河北有序承接北京非首都功能转移和京津科技成果转化。强化区域环保协作，联合开展大气、河流、湖泊等污染治理，加强区域生态屏障建设，共建坝上高原生态防护区、燕山—太行山生态涵养区，推动光伏等新能源广泛应用。创新生态环境联动管理体制机制，构建区域一体化的生态环境监测网络、生态环境信息网络和生态环境应急预警体系，建立区域生态环保协调机制、水资源统一调配制度、跨区域联合监察执法机制，建立健全区域生态保护补偿机制和跨区域排污权交易市场。到2020年，京津冀地区生态环境保护协作机制有效运行，生态环境质量明显改善。

推进长江经济带共抓大保护。把保护和修复长江生态环境摆在首要位置，推进长江经济带生态文明建设，建设水清地绿天蓝的绿色生态廊道。统筹水资源、水环境、水生态，推动上中下游协同发展、东中西部互动合作，加强跨部门、跨区域监管与应急协调联动，把实施重大生态修复工程作为推动长江经济带发展项目的优先选项，共抓大保护，不搞大开发。统筹江河湖泊丰富多样的生态要素，构建以长江干支流为经络，以山水林田湖为有机整体，江湖关系和谐、流域水质优良、生态流量充足、水土保持有效、生物种类多样的生态安全格局。上游区重点加强水源涵养、水土保持功能和生物多样性保护，合理开发利用水资源，严控水电开发生态影响；中游区重点协调江湖关系，确保丹江口水库水质安全；下游区加快产业转型升级，重点加强退化水生态系统恢复，强化饮用水水源保护，严格控制城镇周边生态空间占用，开展河网地区水污染治理。妥善处理江河湖泊关系，实施长江干流及洞庭湖上游"四水"、鄱阳湖上游"五河"的水库群联合调度，保障长江干支流生态流量与两湖生态水位。统筹规划、集约利用长江岸线资源，控制岸线开发强度。强化跨界水质断面考核，推动协同治理。

第四章
深化质量管理，大力实施三大行动计划

以提高环境质量为核心，推进联防联控和流域共治，制定大气、水、土壤三大污染防治行动计划的施工图。根据区域、流域和类型差异分区施策，实施多污染物协同控制，提高治理措施的针对性和有效性。实行环境质量底线管理，努力实现分阶段达到环境质量标准、治理责任清单式落地，解决群众身边的突出环境问题。

第一节　分区施策改善大气环境质量

实施大气环境质量目标管理和限期达标规划。各省（区、市）要对照国家大气环境质量标准，开展形势分析，定期考核并公布大气环境质量信息。强化目标和任务的过程管理，深入推进钢铁、水泥等重污染行业过剩产能退出，大力推进清洁能源使用，推进机动车和油品标准升级，加强油品等能源产品质量监管，加强移动源污染治理，加大城市扬尘和小微企业分散源、生活源污染整治力度。深入实施《大气污染防治行动计划》，大幅削减二氧化硫、氮氧化物和颗粒物的排放量，全面启动挥发性有机物污染防治，开展大气氨排放控制试点，实现全国地级及以上城市二氧化硫、一氧化碳浓度全部达标，细颗粒物、可吸入颗粒物浓度明显下降，二氧化氮浓度继续下降，臭氧浓度保持稳定、力争改善。实施城市大气环境质量目标管理，已经达标的城市，应当加强保护并持续改善；未达标的城市，应确定达标期限，向社会公布，并制定实施限期达标规划，明确达标时间表、路线图和重点任务。

加强重污染天气应对。强化各级空气质量预报中心运行管理，提高预报准确性，及时发布空气质量预报信息，实现预报信息全国共享、联网发布。完善重度及以上污染天气的区域联合预警机制，加强东北、西北、成渝和华中区域大气环境质量预测预报能力。健全应急预案体系，制定重污染天气应急预案实施情况评估技术规程，加强对预案实施情况的检查和评估。各省（区、市）和地级及以上城市及时修编重污染天气应急预案，开展重污染天气成因分析和污染物来源解析，科学制定针对性减排措施，每年更新应急减排措施项目清单。及时启动应急响应措施，提高重污染天气应对的有效性。强化监管和督察，对应对不及时、措施不力的地方政府，视情况予以约谈、通报、挂牌督办。

深化区域大气污染联防联控。全面深化京津冀及周边地区、长三角、珠三角等区域大气污染联防联控，建立常态化区域协作机制，区域内统一规划、统一标准、统一监测、统一防治。对重点行业、领域制定实施统一的环保标准、排污收费政策、能源消费政策，统一老旧车辆淘汰和在用车辆管理标准。重点区域严格控制煤炭消费总量，京津冀及山东、长三角、珠三角等区域，以及空气质量排名较差的前 10 位城市中受燃煤影响较大的城市要实现煤炭消费负增长。通过市场化方式促进老旧车辆、船舶加速淘汰以及防污设施设备改造，强化新生产机动车、非道路移动机械环保达标监管。开展清洁柴油机行动，加强高排放工程机械、重型柴油车、农业机械等管理，重点区域开展柴油车注册登记环保查验，对货运车、客运车、公交车等开展入户环保检查。提高公共车辆中新能源汽车占比，具备条件的城市在 2017 年底前基本实现公交新能源化。落实珠三角、长三角、环渤海京津冀水域船舶排放控制区管理政策，靠港船舶优先使用岸电，建设船舶大气污染物排放遥感监测和油品质量监测网点，开展船舶排放控制区内船舶排放监测和联合监管，构建机动车船和油品环保达标监管体系。加快非道路移动源油品升级。强化城市道路、施工等扬尘监管和城市综合管理。

显著削减京津冀及周边地区颗粒物浓度。以北京市、保定市、廊坊市为重点，突出抓好冬季散煤治理、重点行业综合治理、机动车监管、重污染天气应对，强化高架源的治理和监管，改善区域空气质量。提高接受外输电比例，增加非化石能源供应，重点城市实施天然气替代煤炭工程，推进电力替代煤炭，大幅减少冬季散煤使用量，"十三五"期间，北京、天津、河北、山东、河南五省（市）煤炭消费总量下降 10% 左右。加快区域内机动车排污监控平台建设，重点治理重型柴油车和高排放车辆。到 2020 年，区域细颗粒物污染形势显著好转，臭氧浓度基本稳定。

明显降低长三角区域细颗粒物浓度。加快产业结构调整，依法淘汰能耗、环保等不达标的产能。"十三五"期间，上海、江苏、浙江、安徽四省（市）煤炭消费总量下降 5% 左右，地级及以上城市建成区基本淘汰 35 蒸吨以下燃煤锅炉。全面推进炼油、石化、工业涂装、印刷等行业挥发性有机物综合整治。到 2020 年，长三角区域细颗粒物浓度显著下降，臭氧浓度基本稳定。

大力推动珠三角区域率先实现大气环境质量基本达标。统筹做好细颗粒物和臭氧污染防控，重点抓好挥发性有机物和氮氧化物协同控制。加快区域内产业转型升级，调整和优化能源结构，工业园区与产业聚集区实施集中供热，有条件的发展大型燃气供热锅炉，"十三五"期间，珠三角区域煤炭消费总量下降10%左右。重点推进石化、化工、油品储运销、汽车制造、船舶制造（维修）、集装箱制造、印刷、家具制造、制鞋等行业开展挥发性有机物综合整治。到2020年，实现珠三角区域大气环境质量基本达标，基本消除重度及以上污染天气。

第二节 精准发力提升水环境质量

实施以控制单元为基础的水环境质量目标管理。依据主体功能区规划和行政区划，划定陆域控制单元，建立流域、水生态控制区、水环境控制单元三级分区体系。实施以控制单元为空间基础、以断面水质为管理目标、以排污许可制为核心的流域水环境质量目标管理。优化控制单元水质断面监测网络，建立控制单元产排污与断面水质响应反馈机制，明确划分控制单元水环境质量责任，从严控制污染物排放量。全面推行"河长制"。在黄河、淮河等流域进行试点，分期分批科学确定生态流量（水位），作为流域水量调度的重要参考。深入实施《水污染防治行动计划》，落实控制单元治污责任，完成目标任务。固定污染源排放为主的控制单元，要确定区域、流域重点水污染物和主要超标污染物排放控制目标，实施基于改善水质要求的排污许可，将治污任务逐一落实到控制单元内的各排污单位（含污水处理厂、设有排放口的规模化畜禽养殖单位）。面源（分散源）污染为主或严重缺水的控制单元，要采用政策激励、加强监管以及确保生态基流等措施改善水生态环境。自2017年起，各省份要定期向社会公开控制单元水环境质量目标管理情况。

专栏2　各流域需要改善的控制单元

（一）长江流域（108个）。

双桥河合肥市控制单元等40个单元由Ⅳ类升为Ⅲ类；乌江重庆市控制单元等7个单元由Ⅴ类升为Ⅲ类；来河滁州市控制单元等9个单元由Ⅴ类升为Ⅳ类；京山河荆门市控制单元等2个单元由劣Ⅴ类升为Ⅲ类；沱江内江市控制单元等4个单元由劣Ⅴ类升为Ⅳ类；十五里河合肥市控制单元等24个单元由劣Ⅴ类升为Ⅴ类；滇池外海昆明市控制单元化学需氧量浓度下降；南淝河合肥市控制单元等3个单元氨氮浓度下降；竹皮河荆门市控制单元等4个单元氨氮、总磷浓度下降；岷江宜宾市控制单元等14个单元总磷浓度下降。

（二）海河流域（75个）。

洋河张家口市八号桥控制单元等9个单元由Ⅳ类升为Ⅲ类；妫水河下段北京市控制单元等3个单元由Ⅴ类升为Ⅳ类；潮白河通州区控制单元等26个单元由劣Ⅴ类升为Ⅴ类；宣惠河沧州市控制单元等6个单元化学需氧量浓度下降；通惠河下段北京市控制单元等26个单元氨氮浓度下降；共产主义渠新乡市控制单元等3个单元氨氮、总磷浓度下降；海河天津市海河大闸控制单元化学需氧量、氨氮浓度下降；潮白新河天津市控制单元总磷浓度下降。

(三) 淮河流域 (49个)。

谷河阜阳市控制单元等17个单元由Ⅳ类升为Ⅲ类；东鱼河菏泽市控制单元由Ⅴ类升为Ⅲ类；新濉河宿迁市控制单元等9个单元由Ⅴ类升为Ⅳ类；洙赵新河菏泽市控制单元由劣Ⅴ类升为Ⅲ类；运料河徐州市控制单元由劣Ⅴ类升为Ⅳ类；涡河亳州市岳坊大桥控制单元等16个单元由劣Ⅴ类升为Ⅴ类；包河商丘市控制单元等4个单元氨氮浓度下降。

(四) 黄河流域 (35个)。

伊洛河洛阳市控制单元等14个单元由Ⅳ类升为Ⅲ类；葫芦河固原市控制单元等4个单元由Ⅴ类升为Ⅳ类；岚河吕梁市控制单元由劣Ⅴ类升为Ⅳ类；大黑河乌兰察布市控制单元等8个单元由劣Ⅴ类升为Ⅴ类；昆都仑河包头市控制单元等8个单元氨氮浓度下降。

(五) 松花江流域 (12个)。

小兴凯湖鸡西市控制单元等9个单元由Ⅳ类升为Ⅲ类；阿什河哈尔滨市控制单元由劣Ⅴ类升为Ⅴ类；呼伦湖呼伦贝尔市控制单元化学需氧量浓度下降；饮马河长春市靠山南楼控制单元氨氮浓度下降。

(六) 辽河流域 (13个)。

寇河铁岭市控制单元等6个单元由Ⅳ类升为Ⅲ类；辽河沈阳市巨流河大桥控制单元等3个单元由Ⅴ类升为Ⅳ类；亮子河铁岭市控制单元等2个单元由劣Ⅴ类升为Ⅴ类；浑河抚顺市控制单元总磷浓度下降；条子河四平市控制单元氨氮浓度下降。

（七）珠江流域（17个）。

九洲江湛江市排里控制单元等2个单元由Ⅲ类升为Ⅱ类；潭江江门市牛湾控制单元由Ⅳ类升为Ⅱ类；鉴江茂名市江口门控制单元等4个单元由Ⅳ类升为Ⅲ类；东莞运河东莞市樟村控制单元等2个单元由Ⅴ类升为Ⅳ类；小东江茂名市石碧控制单元由劣Ⅴ类升为Ⅳ类；深圳河深圳市河口控制单元等5个单元由劣Ⅴ类升为Ⅴ类；杞麓湖玉溪市控制单元化学需氧量浓度下降；星云湖玉溪市控制单元总磷浓度下降。

（八）浙闽片河流（25个）。

浦阳江杭州市控制单元等13个单元由Ⅳ类升为Ⅲ类；汀溪厦门市控制单元等3个单元由Ⅴ类升为Ⅲ类；南溪漳州市控制单元等5个单元由Ⅴ类升为Ⅳ类；金清港台州市控制单元等4个单元由劣Ⅴ类升为Ⅴ类。

（九）西北诸河（3个）。

博斯腾湖巴音郭楞蒙古自治州控制单元由Ⅳ类升为Ⅲ类；北大河酒泉市控制单元由劣Ⅴ类升为Ⅲ类；克孜河喀什地区控制单元由劣Ⅴ类升为Ⅴ类。

（十）西南诸河（6个）。

黑惠江大理白族自治州控制单元等4个单元由Ⅳ类升为Ⅲ类；异龙湖红河哈尼族彝族自治州控制单元化学需氧量浓度下降；西洱河大理白族自治州控制单元氨氮浓度下降。

实施流域污染综合治理。实施重点流域水污染防治规划。流域上下游各级政府、各部门之间加强协调配合、定期会商，实施联合监测、联合执法、应急联动、信息共享。长江流域强化系统保护，加大水生生物多样性保护力度，强化水上交通、船舶港口污染防治。实施岷江、沱江、乌江、清水江、长江干流宜昌段总磷污染综合治理，有效控制贵州、四川、湖北、云南等总磷污染。太湖坚持综合治理，增强流域生态系统功能，防范蓝藻暴发，确保饮用水安全；巢湖加强氮、磷总量控制，改善入湖河流水质，修复湖滨生态功能；滇池加强氮、磷总量控制，重点防控城市污水和农业面源污染入湖，分区分步开展生态修复，逐步恢复水生态系统。海河流域突出节水和再生水利用，强化跨界水体治理，重点整治城乡黑臭水体，保障白洋淀、衡水湖、永定河生态需水。淮河流域大幅降低造纸、化肥、酿造等行业污染物排放强度，有效控制氨氮污染，持续改善洪河、涡河、颍河、惠济河、包河等支流水质，切实防控突发污染事件。黄河流域重点控制煤化工、石化企业排放，持续改善汾河、涑水河、总排干、大黑河、乌梁素海、湟水河等支流水质，降低中上游水环境风险。松花江流域持续改善阿什河、伊通河等支流水质，重点解决石化、酿造、制药、造纸等行业污染问题，加大水生态保护力度，进一步增加野生鱼类种群数量，加快恢复湿地生态系统。辽河流域大幅降低石化、造纸、化工、农副食品加工等行业污染物排放强度，持续改善浑河、太子河、条子河、招苏台河等支流水质，显著恢复水生态系统，全面恢复湿地生态系统。珠江流域建立健全广东、广西、云南等联合治污防控体系，重点保障东江、西江供水水质安全，改善珠江三角洲地区水生态环境。

优先保护良好水体。 实施从水源到水龙头全过程监管，持续提升饮用水安全保障水平。地方各级人民政府及供水单位应定期监测、检测和评估本行政区域内饮用水水源、供水厂出水和用户水龙头水质等饮水安全状况。地级及以上城市每季度向社会公开饮水安全状况信息，县级及以上城市自2018年起每季度向社会公开。开展饮用水水源规范化建设，依法清理饮用水水源保护区内违法建筑和排污口。加强农村饮用水水源保护，实施农村饮水安全巩固提升工程。各省（区、市）应于2017年底前，基本完成乡镇及以上集中式饮用水水源保护区划定，开展定期监测和调查评估。到2020年，地级及以上城市集中式饮用水水源水质达到或优于Ⅲ类比例高于93%。对江河源头及现状水质达到或优于Ⅲ类的江河湖库开展生态环境安全评估，制定实施生态环境保护方案，东江、滦河、千岛湖、南四湖等流域于2017年底前完成。七大重点流域制定实施水生生物多样性保护方案。

推进地下水污染综合防治。定期调查评估集中式地下水型饮用水水源补给区和污染源周边区域环境状况。加强重点工业行业地下水环境监管，采取防控措施有效降低地下水污染风险。公布地下水污染地块清单，管控风险，开展地下水污染修复试点。到2020年，全国地下水污染加剧趋势得到初步遏制，质量极差的地下水比例控制在15%左右。

大力整治城市黑臭水体。建立地级及以上城市建成区黑臭水体等污染严重水体清单，制定整治方案，细化分阶段目标和任务安排，向社会公布年度治理进展和水质改善情况。建立全国城市黑臭水体整治监管平台，公布全国黑臭水体清单，接受公众评议。各城市在当地主流媒体公布黑臭水体清单、整治达标期限、责任人、整治进展及效果；建立长效机制，开展水体日常维护与监管工作。2017年底前，直辖市、省会城市、计划单列市建成区基本消除黑臭水体，其他地级城市实现河面无大面积漂浮物、河岸无垃圾、无违法排污口；到2020年，地级及以上城市建成区黑臭水体比例均控制在10%以内，其他城市力争大幅度消除重度黑臭水体。

改善河口和近岸海域生态环境质量。实施近岸海域污染防治方案，加大渤海、东海等近岸海域污染治理力度。强化直排海污染源和沿海工业园区监管，防控沿海地区陆源溢油污染海洋。开展国际航行船舶压载水及污染物治理。规范入海排污口设置，2017年底前，全面清理非法或设置不合理的入海排污口。到2020年，沿海省（区、市）入海河流基本消除劣Ⅴ类的水体。实施蓝色海湾综合治理，重点整治黄河口、长江口、闽江口、珠江口、辽东湾、渤海湾、胶州湾、杭州湾、北部湾等河口海湾污染。严格禁渔休渔措施。控制近海养殖密度，推进生态健康养殖，大力开展水生生物增殖放流，加强人工鱼礁和海洋牧场建设。加强海岸带生态保护与修复，实施"南红北柳"湿地修复工程，严格控制生态敏感地区围填海活动。到2020年，全国自然岸线（不包括海岛岸线）保有率不低于35%，整治修复海岸线1000公里。建设一批海洋自然保护区、海洋特别保护区和水产种质资源保护区，实施生态岛礁工程，加强海洋珍稀物种保护。

第三节　分类防治土壤环境污染

推进基础调查和监测网络建设。全面实施《土壤污染防治行动计划》，以农用地和重点行业企业用地为重点，开展土壤污染状况详查，2018年底前查明农用地土壤污染的面积、分布及其对农产品质量的影响，2020年底前掌握重点行业企业用地中的污染地块分布及其环境风险情况。开展电子废物拆解、废旧塑料回收、非正规垃圾填埋场、历史遗留尾矿库等土壤环境问题集中区域风险排查，建立风险管控名录。统一规划、整合优化土壤环境质量监测点位。充分发挥行业监测网作用，支持各地因地制宜补充增加设置监测点位，增加特征污染物监测项目，提高监测频次。2017年底前，完成土壤环境质量国控监测点位设置，建成国家土壤环境质量监测网络，基本形成土壤环境监测能力；到2020年，实现土壤环境质量监测点位所有县（市、区）全覆盖。

实施农用地土壤环境分类管理。按污染程度将农用地划为三个类别，未污染和轻微污染的划为优先保护类，轻度和中度污染的划为安全利用类，重度污染的划为严格管控类，分别采取相应管理措施。各省级人民政府要对本行政区域内优先保护类耕地面积减少或土壤环境质量下降的县（市、区）进行预警提醒并依法采取环评限批等限制性措施。将符合条件的优先保护类耕地划为永久基本农田，实行严格保护，确保其面积不减少、土壤环境质量不下降。根据土壤污染状况和农产品超标情况，安全利用类耕地集中的县（市、区）要结合当地主要作物品种和种植习惯，制定实施受污染耕地安全利用方案，采取农艺调控、替代种植等措施，降低农产品超标风险。加强对严格管控类耕地的用途管理，依法划定特定农产品禁止生产区域，严禁种植食用农产品，继续在湖南长株潭地区开展重金属污染耕地修复及农作物种植结构调整试点。到2020年，重度污染耕地种植结构调整或退耕还林还草面积力争达到2000万亩。

加强建设用地环境风险管控。建立建设用地土壤环境质量强制调查评估制度。构建土壤环境质量状况、污染地块修复与土地再开发利用协同一体的管理与政策体系。自2017年起，对拟收回土地使用权的有色金属冶炼、石油加工、化工、焦化、电镀、制革等行业企业用地，以及用途拟变更为居住和商业、学校、医疗、养老机构等公共设施的上述企业用地，由土地使用权人负责开展土壤环境状况调查评估；已经收回的，由所在地市、县级人民政府负责开展调查评估。将建设用地土壤环境管理要求纳入城市规划和供地管理，土地开发利用必须符合土壤环境质量要求。暂不开发利用或现阶段不具备治理修复条件的污染地块，由所在地县级人民政府组织划定管控区域，设立标志，发布公告，开展土壤、地表水、地下水、空气环境监测。

开展土壤污染治理与修复。针对典型受污染农用地、污染地块，分批实施200个土壤污染治理与修复技术应用试点项目，加快建立健全技术体系。自2017年起，各地要逐步建立污染地块名录及其开发利用的负面清单，合理确定土地用途。京津冀、长三角、珠三角、东北老工业基地地区城市和矿产资源枯竭型城市等污染地块集中分布的城市，要规范、有序开展再开发利用污染地块治理与修复。长江中下游、成都平原、珠江流域等污染耕地集中分布的省（区、市），应于2018年底前编制实施污染耕地治理与修复方案。2017年底前，发布土壤污染治理与修复责任方终身责任追究办法。建立土壤污染治理与修复全过程监管制度，严格修复方案审查，加强修复过程监督和检查，开展修复成效第三方评估。

强化重点区域土壤污染防治。京津冀区域以城市"退二进三"遗留污染地块为重点，严格管控建设用地开发利用土壤环境风险，加大污灌区、设施农业集中区域土壤环境监测和监管。东北地区加大黑土地保护力度，采取秸秆还田、增施有机肥、轮作休耕等措施实施综合治理。珠江三角洲地区以化工、电镀、印染等重污染行业企业遗留污染地块为重点，强化污染地块开发利用环境监管。湘江流域地区以镉、砷等重金属污染为重点，对污染耕地采取农艺调控、种植结构调整、退耕还林还草等措施，严格控制农产品超标风险。西南地区以有色金属、磷矿等矿产资源开发过程导致的环境污染风险防控为重点，强化磷、汞、铅等历史遗留土壤污染治理。在浙江台州、湖北黄石、湖南常德、广东韶关、广西河池、贵州铜仁等6个地区启动土壤污染综合防治先行区建设。

第五章
实施专项治理，全面推进达标排放与污染减排

以污染源达标排放为底线，以骨干性工程推进为抓手，改革完善总量控制制度，推动行业多污染物协同治污减排，加强城乡统筹治理，严格控制增量，大幅度削减污染物存量，降低生态环境压力。

第一节 实施工业污染源全面达标排放计划

工业污染源全面开展自行监测和信息公开。工业企业要建立环境管理台账制度，开展自行监测，如实申报，属于重点排污单位的还要依法履行信息公开义务。实施排污口规范化整治，2018年底前，工业企业要进一步规范排污口设置，编制年度排污状况报告。排污企业全面实行在线监测，地方各级人民政府要完善重点排污单位污染物超标排放和异常报警机制，逐步实现工业污染源排放监测数据统一采集、公开发布，不断加强社会监督，对企业守法承诺履行情况进行监督检查。2019年底前，建立全国工业

企业环境监管信息平台。

排查并公布未达标工业污染源名单。各地要加强对工业污染源的监督检查，全面推进"双随机"抽查制度，实施环境信用颜色评价，鼓励探索实施企业超标排放计分量化管理。对污染物排放超标或者重点污染物排放超总量的企业予以"黄牌"警示，限制生产或停产整治；对整治后仍不能达到要求且情节严重的企业予以"红牌"处罚，限期停业、关闭。自2017年起，地方各级人民政府要制定本行政区域工业污染源全面达标排放计划，确定年度工作目标，每季度向社会公布"黄牌"、"红牌"企业名单。环境保护部将加大抽查核查力度，对企业超标现象普遍、超标企业集中地区的地方政府进行通报、挂牌督办。

实施重点行业企业达标排放限期改造。建立分行业污染治理实用技术公开遴选与推广应用机制，发布重点行业污染治理技术。分流域分区域制定实施重点行业限期整治方案，升级改造环保设施，加大检查核查力度，确保稳定达标。以钢铁、水泥、石化、有色金属、玻璃、燃煤锅炉、造纸、印染、化工、焦化、氮肥、农副食品加工、原料药制造、制革、农药、电镀等行业为重点，推进行业达标排放改造。

完善工业园区污水集中处理设施。实行"清污分流、雨污分流"，实现废水分类收集、分质处理，入园企业应在达到国家或地方规定的排放标准后接入集中式污水处理设施处理，园区集中式污水处理设施总排口应安装自动监控系统、视频监控系统，并与环境保护主管部门联网。开展工业园区污水集中处理规范化改造示范。

第二节 深入推进重点污染物减排

改革完善总量控制制度。以提高环境质量为核心，以重大减排工程为主要抓手，上下结合，科学确定总量控制要求，实施差别化管理。优化总量减排核算体系，以省级为主体实施核查核算，推动自主减排管理，鼓励将持续有效改善环境质量的措施纳入减排核算。加强对生态环境保护重大工程的调度，对进度滞后地区及早预警通报，各地减排工程、指标情况要主动向社会公开。总量减排考核服从于环境质量考核，重点审查环境质量未达到标准、减排数据与环境质量变化趋势明显不协调的地区，并根据环境保护督查、日常监督检查和排污许可执行情况，对各省（区、市）自主减排管理情况实施"双随机"抽查。大力推行区域性、行业性总量控制，鼓励各地实施特征性污染物总量控制，并纳入各地国民经济和社会发展规划。

推动治污减排工程建设。各省（区、市）要制定实施造纸、印染等十大重点涉水行业专项治理方案，大幅降低污染物排放强度。电力、钢铁、纺织、造纸、石油石化、化工、食品发酵等高耗水行业达到先进定额标准。以燃煤电厂超低排放改造为重点，对电力、钢铁、建材、石化、有色金属等重点行业，实施综合治理，对二氧化硫、氮氧化物、烟粉尘以及重金属等多污染物实施协同控制。各省（区、市）应于2017年底前制定专项治理方案并向社会公开，对治理不到位的工程项目要公开曝光。制定分行业治污技术政策，培育示范企业和示范工程。

第五章 实施专项治理，全面推进达标排放与污染减排

专栏3 推动重点行业治污减排

（一）造纸行业。

力争完成纸浆无元素氯漂白改造或采取其他低污染制浆技术，完善中段水生化处理工艺，增加深度治理工艺，进一步完善中控系统。

（二）印染行业。

实施低排水染整工艺改造及废水综合利用，强化清污分流、分质处理、分质回用，完善中段水生化处理，增加强氧化、膜处理等深度治理工艺。

（三）味精行业。

提高生产废水循环利用水平，分离尾液和离交尾液采用絮凝气浮和蒸发浓缩等措施，外排水采取厌氧—好氧二级生化处理工艺；敏感区域应深度处理。

（四）柠檬酸行业。

采用低浓度废水循环再利用技术，高浓度废水采用喷浆造粒等措施。

（五）氮肥行业。

开展工艺冷凝液水解解析技术改造，实施含氰、含氨废水综合治理。

（六）酒精与啤酒行业。

低浓度废水采用物化—生化工艺，预处理后由园区集中处理。啤酒行业采用就地清洗技术。

（七）制糖行业。

采用无滤布真空吸滤机、高压水清洗、甜菜干法输送及压粕水回收，推进废糖蜜、酒精废醪液发酵还田综合利用，鼓励废水生

化处理后回用，敏感区域执行特别排放限值。

（八）淀粉行业。

采用厌氧＋好氧生化处理技术，建设污水处理设施在线监测和中控系统。

（九）屠宰行业。

强化外排污水预处理，敏感区域执行特别排放限值，有条件的采用膜生物反应器工艺进行深度处理。

（十）磷化工行业。

实施湿法磷酸净化改造，严禁过磷酸钙、钙镁磷肥新增产能。发展磷炉尾气净化合成有机化工产品，鼓励各种建材或建材添加剂综合利用磷渣、磷石膏。

（十一）煤电行业。

加快推进燃煤电厂超低排放和节能改造。强化露天煤场抑尘措施，有条件的实施封闭改造。

（十二）钢铁行业。

完成干熄焦技术改造，不同类型的废水应分别进行预处理。未纳入淘汰计划的烧结机和球团生产设备全部实施全烟气脱硫，禁止设置脱硫设施烟气旁路；烧结机头、机尾、焦炉、高炉出铁场、转炉烟气除尘等设施实施升级改造，露天原料场实施封闭改造，原料转运设施建设封闭皮带通廊，转运站和落料点配套抽风收尘装置。

（十三）建材行业。

原料破碎、生产、运输、装卸等各环节实施堆场及输送设备全封闭、道路清扫等措施，有效控制无组织排放。水泥窑全部实施烟气脱硝，水泥窑及窑磨一体机进行高效除尘改造；平板玻璃行业推进"煤改气"、"煤改电",禁止掺烧高硫石油焦等劣质原料,未使用清洁能源的浮法玻璃生产线全部实施烟气脱硫,浮法玻璃生产线全部实施烟气高效除尘、脱硝;建筑卫生陶瓷行业使用清洁燃料,喷雾干燥塔、陶瓷窑炉安装脱硫除尘设施,氮氧化物不能稳定达标排放的喷雾干燥塔采取脱硝措施。

(十四)石化行业。

催化裂化装置实施催化再生烟气治理,对不能稳定达标排放的硫磺回收尾气,提高硫磺回收率或加装脱硫设施。

(十五)有色金属行业。

加强富余烟气收集,对二氧化硫含量大于3.5%的烟气,采取两转两吸制酸等方式回收。低浓度烟气和制酸尾气排放超标的必须进行脱硫。规范冶炼企业废气排放口设置,取消脱硫设施旁路。

控制重点地区重点行业挥发性有机物排放。全面加强石化、有机化工、表面涂装、包装印刷等重点行业挥发性有机物控制。细颗粒物和臭氧污染严重省份实施行业挥发性有机污染物总量控制,制定挥发性有机污染物总量控制目标和实施方案。强化挥发性有机物与氮氧化物的协同减排,建立固定源、移动源、面源排放清单,对芳香烃、烯烃、炔烃、醛类、酮类等挥发性有机物实施重点减排。开展石化行业"泄漏检测与修复"专项行动,对无组织排放开展治理。各地要明确时限,完成加油站、储油库、油罐车油气回收治理,油气回收率提高到90%以上,并加快推进原油成品油码头油气回收治理。涂装行业实施低挥发性有机物含量涂料替代、涂装工艺与设备改进,建设挥发性有机物收集与治理设施。印刷行业全面开展低挥发性有机物含量原辅料替代,改进生产工艺。京津冀及周边地区、长三角地区、珠三角地区,以及成渝、武汉及其周边、辽宁中部、陕西关中、长株潭等城市群全面加强挥发性有机物排放控制。

总磷、总氮超标水域实施流域、区域性总量控制。总磷超标的控制单元以及上游相关地区要实施总磷总量控制，明确控制指标并作为约束性指标，制定水质达标改善方案。重点开展100家磷矿采选和磷化工企业生产工艺及污水处理设施建设改造。大力推广磷铵生产废水回用，促进磷石膏的综合加工利用，确保磷酸生产企业磷回收率达到96%以上。沿海地级及以上城市和汇入富营养化湖库的河流，实施总氮总量控制，开展总氮污染来源解析，明确重点控制区域、领域和行业，制定总氮总量控制方案，并将总氮纳入区域总量控制指标。氮肥、味精等行业提高辅料利用效率，加大资源回收力度。印染等行业降低尿素的使用量或使用尿素替代助剂。造纸等行业加快废水处理设施精细化管理，严格控制营养盐投加量。强化城镇污水处理厂生物除磷、脱氮工艺，实施畜禽养殖业总磷、总氮与化学需氧量、氨氮协同控制。

专栏4　区域性、流域性总量控制地区

（一）挥发性有机物总量控制。

在细颗粒物和臭氧污染较严重的16个省份实施行业挥发性有机物总量控制，包括：北京市、天津市、河北省、辽宁省、上海市、江苏省、浙江省、安徽省、山东省、河南省、湖北省、湖南省、广东省、重庆市、四川省、陕西省等。

（二）总磷总量控制。

总磷超标的控制单元以及上游相关地区实施总磷总量控制，包括：天津市宝坻区，黑龙江省鸡西市，贵州省黔南布依族苗族自治州、黔东南苗族侗族自治州，河南省漯河市、鹤壁市、安阳市、新乡市，湖北省宜昌市、十堰市，湖南省常德市、益阳市、岳阳市，江西省南昌市、九江市，辽宁省抚顺市，四川省宜宾市、泸州市、眉山市、乐山市、成都市、资阳市，云南省玉溪市等。

（三）总氮总量控制。

在56个沿海地级及以上城市或区域实施总氮总量控制，包括：丹东市、大连市、锦州市、营口市、盘锦市、葫芦岛市、秦皇岛市、唐山市、沧州市、天津市、滨州市、东营市、潍坊市、烟台市、威海市、青岛市、日照市、连云港市、盐城市、南通市、上海市、杭州市、宁波市、温州市、嘉兴市、绍兴市、舟山市、台州市、福州市、平潭综合实验区、厦门市、莆田市、宁德市、漳州市、泉州市、广州市、深圳市、珠海市、汕头市、江门市、湛江市、茂名市、惠州市、汕尾市、阳江市、东莞市、中山市、潮州市、揭阳市、北海市、防城港市、钦州市、海口市、三亚市、三沙市和海南省直辖县级行政区等。

在29个富营养化湖库汇水范围内实施总氮总量控制，包括：安徽省巢湖、南漪湖，安徽省、湖北省龙感湖，北京市怀柔水库，天津市于桥水库，河北省白洋淀，吉林省松花湖，内蒙古自治区呼伦湖、乌梁素海，山东省南四湖，江苏省白马湖、高邮湖、洪泽湖、太湖、阳澄湖，浙江省西湖，上海市、江苏省淀山湖，湖南省洞庭湖，广东省高州水库、鹤地水库，四川省鲁班水库、邛海，云南省滇池、杞麓湖、星云湖、异龙湖，宁夏自治区沙湖、香山湖，新疆自治区艾比湖等。

第三节 加强基础设施建设

加快完善城镇污水处理系统。全面加强城镇污水处理及配套管网建设，加大雨污分流、清污混流污水管网改造，优先推进城中村、老旧城区和城乡结合部污水截流、收集、纳管，消除河水倒灌、地下水渗入等现象。到2020年，全国所有县城和重点镇具备污水收集处理能力，城市和县城污水处理率分别达到95%和85%左右，地级及以上城市建成区基本实现污水全收集、全处理。提升污水再生利用和污泥处置水平，大力推进污泥稳定化、无害化和资源化处理处置，地级及以上城市污泥无害化处理处置率达到90%，京津冀区域达到95%。控制初期雨水污染，排入自然水体的雨水须经过岸线净化，加快建设和改造沿岸截流干管，控制渗漏和合流制污水溢流污染。因地制宜、一河一策，控源截污、内源污染治理多管齐下，科学整治城市黑臭水体；因地制宜实施城镇污水处理厂升级改造，有条件的应配套建设湿地生态处理系统，加强废水资源化、能源化利用。敏感区域（重点湖泊、重点水库、近岸海域汇水区域）城镇污水处理设施应于2017年底前全面达到一级A排放标准。建成区水体水质达不到地表水Ⅳ类标准的城市，新建城镇污水处理设施要执行一级A排放标准。到2020年，实现缺水城市再生水利用率达到20%以上，京津冀区域达到30%以上。将港口、船舶修造厂环卫设施、污水处理设施纳入城市设施建设规划，提升含油污水、化学品洗舱水、生活污水等的处置能力。实施船舶压载水管理。

实现城镇垃圾处理全覆盖和处置设施稳定达标运行。加快县城垃圾处理设施建设，实现城镇垃圾处理设施全覆盖。提高城市生活垃圾处理减量化、资源化和无害化水平,全国城市生活垃圾无害化处理率达到95%以上,90%以上村庄的生活垃圾得到有效治理。大中型城市重点发展生活垃圾焚烧发电技术,鼓励区域共建共享焚烧处理设施,积极发展生物处理技术,合理统筹填埋处理技术,到2020年,垃圾焚烧处理率达到40%。完善收集储运系统,设市城市全面推广密闭化收运,实现干、湿分类收集转运。加强垃圾渗滤液处理处置、焚烧飞灰处理处置、填埋场甲烷利用和恶臭处理,向社会公开垃圾处置设施污染物排放情况。加快建设城市餐厨废弃物、建筑垃圾和废旧纺织品等资源化利用和无害化处理系统。以大中型城市为重点,建设生活垃圾分类示范城市(区)、生活垃圾存量治理示范项目,大中型城市建设餐厨垃圾处理设施。支持水泥窑协同处置城市生活垃圾。

推进海绵城市建设。转变城市规划建设理念,保护和恢复城市生态。老城区以问题为导向,以解决城市内涝、雨水收集利用、黑臭水体治理为突破口,推进区域整体治理,避免大拆大建。城市新区以目标为导向,优先保护生态环境,合理控制开发强度。综合采取"渗、滞、蓄、净、用、排"等措施,加强海绵型建筑与小区、海绵型道路与广场、海绵型公园和绿地、雨水调蓄与排水防涝设施等建设。大力推进城市排水防涝设施的达标建设,加快改造和消除城市易涝点。到2020年,能够将70%的降雨就地消纳和利用的土地面积达到城市建成区面积的20%以上。加强城镇节水,公共建筑必须采用节水器具,鼓励居民家庭选用节水器具。到2020年,地级及以上缺水城市全部达到国家节水型城市标准要求,京津冀、长三角、珠三角等区域提前一年完成。

增加清洁能源供给和使用。优先保障水电和国家"十三五"能源发展相关规划内的风能、太阳能、生物质能等清洁能源项目发电上网，落实可再生能源全额保障性收购政策，到 2020 年，非化石能源装机比重达到 39%。煤炭占能源消费总量的比重降至 58% 以下。扩大城市高污染燃料禁燃区范围，提高城市燃气化率，地级及以上城市供热供气管网覆盖的地区禁止使用散煤，京津冀、长三角、珠三角等重点区域、重点城市实施"煤改气"工程，推进北方地区农村散煤替代。加快城市新能源汽车充电设施建设，政府机关、大中型企事业单位带头配套建设，继续实施新能源汽车推广。

大力推进煤炭清洁化利用。加强商品煤质量管理，限制开发和销售高硫、高灰等煤炭资源，发展煤炭洗选加工，到 2020 年，煤炭入洗率提高到 75% 以上。大力推进以电代煤、以气代煤和以其他清洁能源代煤，对暂不具备煤炭改清洁燃料条件的地区，积极推进洁净煤替代。建设洁净煤配送中心，建立以县（区）为单位的全密闭配煤中心以及覆盖所有乡镇、村的洁净煤供应网络。加快纯凝（只发电不供热）发电机组供热改造，鼓励热电联产机组替代燃煤小锅炉，推进城市集中供热。到 2017 年，除确有必要保留的外，全国地级及以上城市建成区基本淘汰 10 蒸吨以下燃煤锅炉。

第四节　加快农业农村环境综合治理

继续推进农村环境综合整治。继续深入开展爱国卫生运动，持续推进城乡环境卫生整治行动，建设健康、宜居、美丽家园。深化"以奖促治"政策，以南水北调沿线、三峡库区、长江沿线等重要水源地周边为重点，推进新一轮农村环境连片整治，有条件的省份开展全覆盖拉网式整治。因地制宜开展治理，完善农村生活垃圾"村收集、镇转运、县处理"模式，鼓励就地资源化，加快整治"垃圾围村"、"垃圾围坝"等问题，切实防止城镇垃圾向农村转移。整县推进农村污水处理统一规划、建设、管理。积极推进城镇污水、垃圾处理设施和服务向农村延伸，开展农村厕所无害化改造。继续实施农村清洁工程，开展河道清淤疏浚。到2020年，新增完成环境综合整治建制村13万个。

大力推进畜禽养殖污染防治。划定禁止建设畜禽规模养殖场（小区）区域，加强分区分类管理，以废弃物资源化利用为途径，整县推进畜禽养殖污染防治。养殖密集区推行粪污集中处理和资源化综合利用。2017年底前，各地区依法关闭或搬迁禁养区内的畜禽养殖场（小区）和养殖专业户。大力支持畜禽规模养殖场（小区）标准化改造和建设。

打好农业面源污染治理攻坚战。优化调整农业结构和布局，推广资源节约型农业清洁生产技术，推动资源节约型、环境友好型、生态保育型农业发展。建设生态沟渠、污水净化塘、地表径流集蓄池等设施，净化农田排水及地表径流。实施环水有机农业行动计划。推进健康生态养殖。实行测土配方施肥。推进种植业清洁生产，开展农膜回收利用，率先实现东北黑土地大田生产地膜零增长。在环渤海京津冀、长三角、珠三角等重点区域，开展种植业和养殖业重点排放源氨防控研究与示范。研究建立农药使用环境影响后评价制度，制定农药包装废弃物回收处理办法。到2020年，实现化肥农药使用量零增长，化肥利用率提高到40%以上，农膜回收率达到80%以上；京津冀、长三角、珠三角等区域提前一年完成。

强化秸秆综合利用与禁烧。建立逐级监督落实机制，疏堵结合、以疏为主，完善秸秆收储体系，支持秸秆代木、纤维原料、清洁制浆、生物质能、商品有机肥等新技术产业化发展，加快推进秸秆综合利用；强化重点区域和重点时段秸秆禁烧措施，不断提高禁烧监管水平。

第六章
实行全程管控，有效防范和降低环境风险

提升风险防控基础能力，将风险纳入常态化管理，系统构建事前严防、事中严管、事后处置的全过程、多层级风险防范体系，严密防控重金属、危险废物、有毒有害化学品、核与辐射等重点领域环境风险，强化核与辐射安全监管体系和能力建设，有效控制影响健康的生态和社会环境危险因素，守牢安全底线。

第一节 完善风险防控和应急响应体系

加强风险评估与源头防控。完善企业突发环境事件风险评估制度，推进突发环境事件风险分类分级管理，严格重大突发环境事件风险企业监管。改进危险废物鉴别体系。选择典型区域、工业园区、流域开展试点，进行废水综合毒性评估、区域突发环境事件风险评估，以此作为行业准入、产业布局与结构调整的基本依据，发布典型区域环境风险评估报告范例。

开展环境与健康调查、监测和风险评估。制定环境与健康工

作办法，建立环境与健康调查、监测和风险评估制度，形成配套政策、标准和技术体系。开展重点地区、流域、行业环境与健康调查，初步建立环境健康风险哨点监测工作网络，识别和评估重点地区、流域、行业的环境健康风险，对造成环境健康风险的企业和污染物实施清单管理，研究发布一批利于人体健康的环境基准。

严格环境风险预警管理。强化重污染天气、饮用水水源地、有毒有害气体、核安全等预警工作，开展饮用水水源地水质生物毒性、化工园区有毒有害气体等监测预警试点。

强化突发环境事件应急处置管理。健全国家、省、市、县四级联动的突发环境事件应急管理体系，深入推进跨区域、跨部门的突发环境事件应急协调机制，健全综合应急救援体系，建立社会化应急救援机制。完善突发环境事件现场指挥与协调制度，以及信息报告和公开机制。加强突发环境事件调查、突发环境事件环境影响和损失评估制度建设。

加强风险防控基础能力。构建生产、运输、贮存、处置环节的环境风险监测预警网络，建设"能定位、能查询、能跟踪、能预警、能考核"的危险废物全过程信息化监管体系。建立健全突发环境事件应急指挥决策支持系统，完善环境风险源、敏感目标、环境应急能力及环境应急预案等数据库。加强石化等重点行业以及政府和部门突发环境事件应急预案管理。建设国家环境应急救援实训基地，加强环境应急管理队伍、专家队伍建设，强化环境应急物资储备和信息化建设，增强应急监测能力。推动环境应急装备产业化、社会化，推进环境应急能力标准化建设。

第二节 加大重金属污染防治力度

加强重点行业环境管理。严格控制涉重金属新增产能快速扩张，优化产业布局，继续淘汰涉重金属重点行业落后产能。涉重金属行业分布集中、产业规模大、发展速度快、环境问题突出的地区，制定实施更严格的地方污染物排放标准和环境准入标准，依法关停达标无望、治理整顿后仍不能稳定达标的涉重金属企业。制定电镀、制革、铅蓄电池等行业工业园区综合整治方案，推动园区清洁、规范发展。强化涉重金属工业园区和重点工矿企业的重金属污染物排放及周边环境中的重金属监测，加强环境风险隐患排查，向社会公开涉重金属企业生产排放、环境管理和环境质量等信息。组织开展金属矿采选冶炼、钢铁等典型行业和贵州黔西南布依族苗族自治州等典型地区铊污染排放调查，制定铊污染防治方案。加强进口矿产品中重金属等环保项目质量监管。

深化重点区域分类防控。重金属污染防控重点区域制定实施重金属污染综合防治规划，有效防控环境风险和改善区域环境质量，分区指导、一区一策，实施差别化防控管理，加快湘江等流域、区域突出问题综合整治，"十三五"期间，争取20个左右地区退出重点区域。在江苏靖江市、浙江平阳县等16个重点区域和江西大余县浮江河流域等8个流域开展重金属污染综合整治示范，探索建立区域和流域重金属污染治理与风险防控的技术和管理体系。建立"锰三角"（锰矿开采和生产过程中存在严重环境污染问题的重庆市秀山县、湖南省花垣县、贵州省松桃县三个县）综合防控协调机制，统一制定综合整治规划。优化调整重点区域环境质量监测点位，2018年底前建成全国重金属环境监测体系。

专栏5　重金属综合整治示范

（一）区域综合防控（16个）。

泰州靖江市（电镀行业综合整治）、温州平阳县（产业入园升级与综合整治）、湖州长兴县（铅蓄电池行业综合整治）、济源市（重金属综合治理与环境监测）、黄石大冶市及周边地区（铜冶炼治理与历史遗留污染整治）、湘潭竹埠港及周边地区（历史遗留污染治理）、衡阳水口山及周边地区（行业综合整治提升）、郴州三十六湾及周边地区（历史遗留污染整治和环境风险预警监控）、常德石门县雄黄矿地区（历史遗留砷污染治理与风险防控）、河池金城江区（结构调整与历史遗留污染整治）、重庆秀山县（电解锰行业综合治理）、凉山西昌市（有色行业整治及污染地块治理）、铜仁万山区（汞污染综合整治）、红河个旧市（产业调整与历史遗留污染整治）、渭南潼关县（有色行业综合整治）、金昌市金川区（产业升级与历史遗留综合整治）。

（二）流域综合整治（8个）。

赣州大余县浮江河流域（砷）、三门峡灵宝市宏农涧河流域（镉、汞）、荆门钟祥市利河—南泉河流域（砷）、韶关大宝山矿区横石水流域（镉）、河池市南丹县刁江流域（砷、镉）、黔南独山县都柳江流域（锑）、怒江兰坪县沘江流域（铅、镉）、陇南徽县永宁河流域（铅、砷）。

加强汞污染控制。禁止新建采用含汞工艺的电石法聚氯乙烯生产项目，到2020年聚氯乙烯行业每单位产品用汞量在2010年的基础上减少50%。加强燃煤电厂等重点行业汞污染排放控制。禁止新建原生汞矿，逐步停止原生汞开采。淘汰含汞体温计、血压计等添汞产品。

第三节　提高危险废物处置水平

合理配置危险废物安全处置能力。各省（区、市）应组织开展危险废物产生、利用处置能力和设施运行情况评估，科学规划并实施危险废物集中处置设施建设规划，将危险废物集中处置设施纳入当地公共基础设施统筹建设。鼓励大型石油化工等产业基地配套建设危险废物利用处置设施。鼓励产生量大、种类单一的企业和园区配套建设危险废物收集贮存、预处理和处置设施，引导和规范水泥窑协同处置危险废物。开展典型危险废物集中处置设施累积性环境风险评价与防控，淘汰一批工艺落后、不符合标准规范的设施，提标改造一批设施，规范管理一批设施。

防控危险废物环境风险。动态修订国家危险废物名录，开展全国危险废物普查，2020年底前，力争基本摸清全国重点行业危险废物产生、贮存、利用和处置状况。以石化和化工行业为重点，打击危险废物非法转移和利用处置违法犯罪活动。加强进口石化和化工产品质量安全监管，打击以原油、燃料油、润滑油等产品名义进口废油等固体废物。继续开展危险废物规范化管理督查考核，以含铬、铅、汞、镉、砷等重金属废物和生活垃圾焚烧飞灰、抗生素菌渣、高毒持久性废物等为重点开展专项整治。制定废铅蓄电池回收管理办法。明确危险废物利用处置二次污染控制要求及综合利用过程环境保护要求，制定综合利用产品中有毒有害物质含量限值，促进危险废物安全利用。

推进医疗废物安全处置。扩大医疗废物集中处置设施服务范围，建立区域医疗废物协同与应急处置机制，因地制宜推进农村、乡镇和偏远地区医疗废物安全处置。实施医疗废物焚烧设施提标改造工程。提高规范化管理水平，严厉打击医疗废物非法买卖等行为，建立医疗废物特许经营退出机制，严格落实医疗废物处置收费政策。

第四节 夯实化学品风险防控基础

评估现有化学品环境和健康风险。开展一批现有化学品危害初步筛查和风险评估，评估化学品在环境中的积累和风险情况。2017年底前，公布优先控制化学品名录，严格限制高风险化学品生产、使用、进口，并逐步淘汰替代。加强有毒有害化学品环境与健康风险评估能力建设。

削减淘汰公约管制化学品。到2020年，基本淘汰林丹、全氟辛基磺酸及其盐类和全氟辛基磺酰氟、硫丹等一批《关于持久性有机污染物的斯德哥尔摩公约》管制的化学品。强化对拟限制或禁止的持久性有机污染物替代品、最佳可行技术以及相关监测检测设备的研发。

严格控制环境激素类化学品污染。2017年底前，完成环境激素类化学品生产使用情况调查，监控、评估水源地、农产品种植区及水产品集中养殖区风险，实行环境激素类化学品淘汰、限制、替代等措施。

第五节 加强核与辐射安全管理

我国是核能核技术利用大国。"十三五"期间，要强化核安全监管体系和监管能力建设，加快推进核安全法治进程，落实核安全规划，依法从严监管，严防发生放射性污染环境的核事故。

提高核设施、放射源安全水平。持续提高核电厂安全运行水平，加强在建核电机组质量监督，确保新建核电厂满足国际最新核安全标准。加快研究堆、核燃料循环设施安全改进。优化核安全设备许可管理，提高核安全设备质量和可靠性。实施加强放射源安全行动计划。

推进放射性污染防治。加快老旧核设施退役和放射性废物处理处置，进一步提升放射性废物处理处置能力，落实废物最小化政策。推进铀矿冶设施退役治理和环境恢复，加强铀矿冶和伴生放射性矿监督管理。

强化核与辐射安全监管体系和能力建设。加强核与辐射安全监管体制机制建设，将核安全关键技术纳入国家重点研发计划。强化国家、区域、省级核事故应急物资储备和能力建设。建成国家核与辐射安全监管技术研发基地。建立国家核安全监控预警和应急响应平台，完善全国辐射环境监测网络，加强国家、省、地市级核与辐射安全监管能力。

第七章
加大保护力度，强化生态修复

贯彻"山水林田湖是一个生命共同体"理念，坚持保护优先、自然恢复为主，推进重点区域和重要生态系统保护与修复，构建生态廊道和生物多样性保护网络，全面提升各类生态系统稳定性和生态服务功能，筑牢生态安全屏障。

第一节 维护国家生态安全

系统维护国家生态安全。识别事关国家生态安全的重要区域，以生态安全屏障以及大江大河重要水系为骨架，以国家重点生态功能区为支撑，以国家禁止开发区域为节点，以生态廊道和生物多样性保护网络为脉络，优先加强生态保护，维护国家生态安全。

建设"两屏三带"国家生态安全屏障。建设青藏高原生态安全屏障，推进青藏高原区域生态建设与环境保护，重点保护好多样、独特的生态系统。推进黄土高原—川滇生态安全屏障建设，重点加强水土流失防治和天然植被保护，保障长江、黄河中下游地区生态安全。建设东北森林带生态安全屏障，重点保护好森林资源和生物多样性，维护东北平原生态安全。建设北方防沙带生态安全屏障，重点加强防护林建设、草原保护和防风固沙，对暂不具备治理条件的沙化土地实行封禁保护，保障"三北"地区生态安全。建设南方丘陵山地带生态安全屏障，重点加强植被修复和水土流失防治，保障华南和西南地区生态安全。

构建生物多样性保护网络。深入实施中国生物多样性保护战略与行动计划，继续开展联合国生物多样性十年中国行动，编制实施地方生物多样性保护行动计划。加强生物多样性保护优先区域管理，构建生物多样性保护网络，完善生物多样性迁地保护设施，实现对生物多样性的系统保护。开展生物多样性与生态系统服务价值评估与示范。

第二节　管护重点生态区域

深化国家重点生态功能区保护和管理。制定国家重点生态功能区产业准入负面清单，制定区域限制和禁止发展的产业目录。优化转移支付政策，强化对区域生态功能稳定性和提供生态产品能力的评价和考核。支持甘肃生态安全屏障综合示范区建设，推进沿黄生态经济带建设。加快重点生态功能区生态保护与建设项目实施，加强对开发建设活动的生态监管，保护区域内重点野生动植物资源，明显提升重点生态功能区生态系统服务功能。

优先加强自然保护区建设与管理。优化自然保护区布局，将重要河湖、海洋、草原生态系统及水生生物、自然遗迹、极小种群野生植物和极度濒危野生动物的保护空缺作为新建自然保护区重点，建设自然保护区群和保护小区，全面提高自然保护区管理系统化、精细化、信息化水平。建立全国自然保护区"天地一体化"动态监测体系，利用遥感等手段开展监测，国家级自然保护区每年监测两次，省级自然保护区每年监测一次。定期组织自然保护区专项执法检查，严肃查处违法违规活动，加强问责监督。加强自然保护区综合科学考察、基础调查和管理评估。积极推进全国自然保护区范围界限核准和勘界立标工作，开展自然保护区土地确权和用途管制，有步骤地对居住在自然保护区核心区和缓冲区的居民实施生态移民。到2020年，全国自然保护区陆地面积占我国陆地国土面积的比例稳定在15%左右，国家重点保护野生动植物种类和典型生态系统类型得到保护的占90%以上。

整合设立一批国家公园。加强对国家公园试点的指导，在试点基础上研究制定建立国家公园体制总体方案。合理界定国家公园范围，整合完善分类科学、保护有力的自然保护地体系，更好地保护自然生态和自然文化遗产原真性、完整性。加强风景名胜区、自然文化遗产、森林公园、沙漠公园、地质公园等各类保护地规划、建设和管理的统筹协调，提高保护管理效能。

第三节 保护重要生态系统

保护森林生态系统。完善天然林保护制度，强化天然林保护和抚育，健全和落实天然林管护体系，加强管护基础设施建设，实现管护区域全覆盖，全面停止天然林商业性采伐。继续实施森林管护和培育、公益林建设补助政策。严格保护林地资源，分级分类进行林地用途管制。到2020年，林地保有量达到31230万公顷。

推进森林质量精准提升。坚持保护优先、自然恢复为主，坚持数量和质量并重、质量优先，坚持封山育林、人工造林并举，宜封则封、宜造则造、宜林则林、宜灌则灌、宜草则草，强化森林经营，大力培育混交林，推进退化林修复，优化森林组成、结构和功能。到 2020 年，混交林占比达到 45%，单位面积森林蓄积量达到 95 立方米/公顷，森林植被碳储量达到 95 亿吨。

保护草原生态系统。稳定和完善草原承包经营制度，实行基本草原保护制度，落实草畜平衡、禁牧休牧和划区轮牧等制度。严格草原用途管制，加强草原管护员队伍建设，严厉打击非法征占用草原、开垦草原、乱采滥挖草原野生植物等破坏草原的违法犯罪行为。开展草原资源调查和统计，建立草原生产、生态监测预警系统。加强"三化"草原治理，防治鼠虫草害。到 2020 年，治理"三化"草原 3000 万公顷。

保护湿地生态系统。开展湿地生态效益补偿试点、退耕还湿试点。在国际和国家重要湿地、湿地自然保护区、国家湿地公园，实施湿地保护与修复工程，逐步恢复湿地生态功能，扩大湿地面积。提升湿地保护与管理能力。

第四节 提升生态系统功能

大规模绿化国土。开展大规模国土绿化行动，加强农田林网建设，建设配置合理、结构稳定、功能完善的城乡绿地，形成沿海、沿江、沿线、沿边、沿湖（库）、沿岛的国土绿化网格，促进山脉、平原、河湖、城市、乡村绿化协同。

继续实施新一轮退耕还林还草和退牧还草。扩大新一轮退耕还林还草范围和规模，在具备条件的 25 度以上坡耕地、严重沙化耕地和重要水源地 15—25 度坡耕地实施退耕还林还草。实施全国退牧还草工程建设规划，稳定扩大退牧还草范围，转变草原畜牧业生产方式，建设草原保护基础设施，保护和改善天然草原生态。

建设防护林体系。加强"三北"、长江、珠江、太行山、沿海等防护林体系建设。"三北"地区乔灌草相结合，突出重点、规模治理、整体推进。长江流域推进退化林修复，提高森林质量，构建"两湖一库"防护林体系。珠江流域推进退化林修复。太行山脉优化林分结构。沿海地区推进海岸基干林带和消浪林建设，修复退化林，完善沿海防护林体系和防灾减灾体系。在粮食主产区营造农田林网，加强村镇绿化，提高平原农区防护林体系综合功能。

建设储备林。在水土光热条件较好的南方省区和其他适宜地区，吸引社会资本参与储备林投资、运营和管理，加快推进储备林建设。在东北、内蒙古等重点国有林区，采取人工林集约栽培、现有林改培、抚育及补植补造等措施，建设以用材林和珍贵树种培育为主体的储备林基地。到2020年，建设储备林1400万公顷，每年新增木材供应能力9500万立方米以上。

培育国土绿化新机制。继续坚持全国动员、全民动手、全社会搞绿化的指导方针，鼓励家庭林场、林业专业合作组织、企业、社会组织、个人开展专业化规模化造林绿化。发挥国有林区和林场在绿化国土中的带动作用，开展多种形式的场外合作造林和森林保育经营，鼓励国有林场担负区域国土绿化和生态修复主体任务。创新产权模式，鼓励地方探索在重要生态区域通过赎买、置换等方式调整商品林为公益林的政策。

第五节 修复生态退化地区

综合治理水土流失。加强长江中上游、黄河中上游、西南岩溶区、东北黑土区等重点区域水土保持工程建设，加强黄土高原地区沟壑区固沟保塬工作，推进东北黑土区侵蚀沟治理，加快南方丘陵地带崩岗治理，积极开展生态清洁小流域建设。

推进荒漠化石漠化治理。加快实施全国防沙治沙规划，开展固沙治沙，加大对主要风沙源区、风沙口、沙尘路径区、沙化扩展活跃区等治理力度，加强"一带一路"沿线防沙治沙，推进沙化土地封禁保护区和防沙治沙综合示范区建设。继续实施京津风沙源治理二期工程，进一步遏制沙尘危害。以"一片两江"（滇桂黔石漠化片区和长江、珠江）岩溶地区为重点，开展石漠化综合治理。到2020年，努力建成10个百万亩、100个十万亩、1000个万亩防沙治沙基地。

加强矿山地质环境保护与生态恢复。严格实施矿产资源开发环境影响评价，建设绿色矿山。加大矿山植被恢复和地质环境综合治理，开展病危险尾矿库和"头顶库"（1公里内有居民或重要设施的尾矿库）专项整治，强化历史遗留矿山地质环境恢复和综合治理。推广实施尾矿库充填开采等技术，建设一批"无尾矿山"（通过有效手段实现无尾矿或仅有少量尾矿占地堆存的矿山），推进工矿废弃地修复利用。

第六节　扩大生态产品供给

推进绿色产业建设。加强林业资源基地建设，加快产业转型升级，促进产业高端化、品牌化、特色化、定制化，满足人民群众对优质绿色产品的需求。建设一批具有影响力的花卉苗木示范基地，发展一批增收带动能力强的木本粮油、特色经济林、林下经济、林业生物产业、沙产业、野生动物驯养繁殖利用示范基地。

加快发展和提升森林旅游休闲康养、湿地度假、沙漠探秘、野生动物观赏等产业，加快林产工业、林业装备制造业技术改造和创新，打造一批竞争力强、特色鲜明的产业集群和示范园区，建立绿色产业和全国重点林产品市场监测预警体系。

构建生态公共服务网络。加大自然保护地、生态体验地的公共服务设施建设力度，开发和提供优质的生态教育、游憩休闲、健康养生养老等生态服务产品。加快建设生态标志系统、绿道网络、环卫、安全等公共服务设施，精心设计打造以森林、湿地、沙漠、野生动植物栖息地、花卉苗木为景观依托的生态体验精品旅游线路，集中建设一批公共营地、生态驿站，提高生态体验产品档次和服务水平。

加强风景名胜区和世界遗产保护与管理。开展风景名胜区资源普查，稳步做好世界自然遗产、自然与文化双遗产培育与申报。强化风景名胜区和世界遗产的管理，实施遥感动态监测，严格控制利用方式和强度。加大保护投入，加强风景名胜区保护利用设施建设。

维护修复城市自然生态系统。提高城市生物多样性，加强城市绿地保护，完善城市绿线管理。优化城市绿地布局，建设绿道绿廊，使城市森林、绿地、水系、河湖、耕地形成完整的生态网络。扩大绿地、水域等生态空间，合理规划建设各类城市绿地，推广立体绿化、屋顶绿化。开展城市山体、水体、废弃地、绿地修复，通过自然恢复和人工修复相结合的措施，实施城市生态修复示范工程项目。加强城市周边和城市群绿化，实施"退工还林"，成片建设城市森林。大力提高建成区绿化覆盖率，加快老旧公园改造，提升公园绿地服务功能。推行生态绿化方式，广植当地树种，乔灌草合理搭配、自然生长。加强古树名木保护，严禁移植天然大树进城。发展森林城市、园林城市、森林小镇。到2020年，城市人均公园绿地面积达到14.6平方米，城市建成区绿地率达到38.9%。

第七节　保护生物多样性

开展生物多样性本底调查和观测。实施生物多样性保护重大工程，以生物多样性保护优先区域为重点，开展生态系统、物种、遗传资源及相关传统知识调查与评估，建立全国生物多样性数据库和信息平台。到2020年，基本摸清生物多样性保护优先区域本底状况。完善生物多样性观测体系，开展生物多样性综合观测站和观测样区建设。对重要生物类群和生态系统、国家重点保护物种及其栖息地开展常态化观测、监测、评价和预警。

实施濒危野生动植物抢救性保护。保护、修复和扩大珍稀濒危野生动植物栖息地、原生境保护区（点），优先实施重点保护野生动物和极小种群野生植物保护工程，开发濒危物种繁育、恢复和保护技术，加强珍稀濒危野生动植物救护、繁育和野化放归，开展长江经济带及重点流域人工种群野化放归试点示范，科学进行珍稀濒危野生动植物再引入。优化全国野生动物救护网络，完善布局并建设一批野生动物救护繁育中心，建设兰科植物等珍稀濒危植物的人工繁育中心。强化野生动植物及其制品利用监管，开展野生动植物繁育利用及其制品的认证标识。调整修订国家重点保护野生动植物名录。

加强生物遗传资源保护。建立生物遗传资源及相关传统知识获取与惠益分享制度，规范生物遗传资源采集、保存、交换、合作研究和开发利用活动，加强与遗传资源相关传统知识保护。开展生物遗传资源价值评估，加强对生物资源的发掘、整理、检测、培育和性状评价，筛选优良生物遗传基因。强化野生动植物基因保护，建设野生动植物人工种群保育基地和基因库。完善西南部生物遗传资源库，新建中东部生物遗传资源库，收集保存国家特有、珍稀濒危及具有重要价值的生物遗传资源。建设药用植物资源、农作物种质资源、野生花卉种质资源、林木种质资源中长期保存库（圃），合理规划和建设植物园、动物园、野生动物繁育中心。

强化野生动植物进出口管理。加强生物遗传资源、野生动植物及其制品进出口管理，建立部门信息共享、联防联控的工作机制，建立和完善进出口电子信息网络系统。严厉打击象牙等野生动植物制品非法交易，构建情报信息分析研究和共享平台，组建打击非法交易犯罪合作机制，严控特有、珍稀、濒危野生动植物种质资源流失。

防范生物安全风险。加强对野生动植物疫病的防护。建立健全国家生态安全动态监测预警体系，定期对生态风险开展全面调查评估。加强转基因生物环境释放监管，开展转基因生物环境释放风险评价和跟踪监测。建设国门生物安全保护网，完善国门生物安全查验机制，严格外来物种引入管理。严防严控外来有害生物物种入侵，开展外来入侵物种普查、监测与生态影响评价，对造成重大生态危害的外来入侵物种开展治理和清除。

第八章
加快制度创新，积极推进治理体系和治理能力现代化

统筹推进生态环境治理体系建设，以环保督察巡视、编制自然资源资产负债表、领导干部自然资源资产离任审计、生态环境损害责任追究等落实地方环境保护责任，以环境司法、排污许可、损害赔偿等落实企业主体责任，加强信息公开，推进公益诉讼，强化绿色金融等市场激励机制，形成政府、企业、公众共治的治理体系。

第一节　健全法治体系

完善法律法规。积极推进资源环境类法律法规制修订。适时完善水污染防治、环境噪声污染防治、土壤污染防治、生态保护补偿、自然保护区等相关制度。

严格环境执法监督。完善环境执法监督机制，推进联合执法、区域执法、交叉执法，强化执法监督和责任追究。进一步明确环

境执法部门行政调查、行政处罚、行政强制等职责,有序整合不同领域、不同部门、不同层次的执法监督力量,推动环境执法力量向基层延伸。

推进环境司法。健全行政执法和环境司法的衔接机制,完善程序衔接、案件移送、申请强制执行等方面规定,加强环保部门与公安机关、人民检察院和人民法院的沟通协调。健全环境案件审理制度。积极配合司法机关做好相关司法解释的制修订工作。

第二节 完善市场机制

推行排污权交易制度。建立健全排污权初始分配和交易制度,落实排污权有偿使用制度,推进排污权有偿使用和交易试点,加强排污权交易平台建设。鼓励新建项目污染物排放指标通过交易方式取得,且不得增加本地区污染物排放总量。推行用能预算管理制度,开展用能权有偿使用和交易试点。

发挥财政税收政策引导作用。开征环境保护税。全面推进资源税改革,逐步将资源税扩展到占用各种自然生态空间范畴。落实环境保护、生态建设、新能源开发利用的税收优惠政策。研究制定重点危险废物集中处置设施、场所的退役费用预提政策。

深化资源环境价格改革。完善资源环境价格机制,全面反映市场供求、资源稀缺程度、生态环境损害成本和修复效益等因素。落实调整污水处理费和水资源费征收标准政策,提高垃圾处理费收缴率,完善再生水价格机制。研究完善燃煤电厂环保电价政策,加大高耗能、高耗水、高污染行业差别化电价水价等政策实施力度。

加快环境治理市场主体培育。探索环境治理项目与经营开发项目组合开发模式，健全社会资本投资环境治理回报机制。深化环境服务试点，创新区域环境治理一体化、环保"互联网+"、环保物联网等污染治理与管理模式，鼓励各类投资进入环保市场。废止各类妨碍形成全国统一市场和公平竞争的制度规定，加强环境治理市场信用体系建设，规范市场环境。鼓励推行环境治理依效付费与环境绩效合同服务。

建立绿色金融体系。建立绿色评级体系以及公益性的环境成本核算和影响评估体系，明确贷款人尽职免责要求和环境保护法律责任。鼓励各类金融机构加大绿色信贷发放力度。在环境高风险领域建立环境污染强制责任保险制度。研究设立绿色股票指数和发展相关投资产品。鼓励银行和企业发行绿色债券，鼓励对绿色信贷资产实行证券化。加大风险补偿力度，支持开展排污权、收费权、购买服务协议抵押等担保贷款业务。支持设立市场化运作的各类绿色发展基金。

加快建立多元化生态保护补偿机制。加大对重点生态功能区的转移支付力度，合理提高补偿标准，向生态敏感和脆弱地区、流域倾斜，推进有关转移支付分配与生态保护成效挂钩，探索资金、政策、产业及技术等多元互补方式。完善补偿范围，逐步实现森林、草原、湿地、荒漠、河流、海洋和耕地等重点领域和禁止开发区域、重点生态功能区等重要区域全覆盖。中央财政支持引导建立跨省域的生态受益地区和保护地区、流域上游与下游的横向补偿机制，推进省级区域内横向补偿。在长江、黄河等重要河流探索开展横向生态保护补偿试点。深入推进南水北调中线工程水源区对口支援、新安江水环境生态补偿试点，推动在京津冀水源涵养区、广西广东九洲江、福建广东汀江—韩江、江西广东东江、云南贵州广西广东西江等开展跨地区生态保护补偿试点。

到 2017 年，建立京津冀区域生态保护补偿机制，将北京、天津支持河北开展生态建设与环境保护制度化。

第三节　落实地方责任

落实政府生态环境保护责任。建立健全职责明晰、分工合理的环境保护责任体系，加强监督检查，推动落实环境保护党政同责、一岗双责。省级人民政府对本行政区域生态环境和资源保护负总责，对区域流域生态环保负相应责任，统筹推进区域环境基本公共服务均等化，市级人民政府强化统筹和综合管理职责，区县人民政府负责执行落实。

改革生态环境保护体制机制。积极推行省以下环保机构监测监察执法垂直管理制度改革试点，加强对地方政府及其相关部门环保履责情况的监督检查。建立区域流域联防联控和城乡协同的治理模式。建立和完善严格监管所有污染物排放的环境保护管理制度。

推进战略和规划环评。在完成京津冀、长三角、珠三角地区及长江经济带、"一带一路"战略环评基础上，稳步推进省、市两级行政区战略环评。探索开展重大政策环境影响论证试点。严格开展开发建设规划环评，作为规划编制、审批、实施的重要依据。深入开展城市、新区总体规划环评，强化规划环评生态空间保护，完善规划环评会商机制。以产业园区规划环评为重点，推进空间和环境准入的清单管理，探索园区内建设项目环评审批管理改革。加强项目环评与规划环评联动，建设四级环保部门环评审批信息联网系统。地方政府和有关部门要依据战略、规划环评，把空间管制、总量管控和环境准入等要求转化为区域开发和保护的刚性约束。严格规划环评责任追究，加强对地方政府和有关部门规划环评工作开展情况的监督。

编制自然资源资产负债表。探索编制自然资源资产负债表，建立实物量核算账户，建立生态环境价值评估制度，开展生态环境资产清查与核算。实行领导干部自然资源资产离任审计，推动地方领导干部落实自然资源资产管理责任。在完成编制自然资源资产负债表试点基础上，逐步建立健全自然资源资产负债表编制制度，在国家层面探索形成主要自然资源资产价值量核算技术方法。

建立资源环境承载能力监测预警机制。研究制定监测评价、预警指标体系和技术方法，开展资源环境承载能力监测预警与成因解析，对资源消耗和环境容量接近或超过承载能力的地区实行预警提醒和差异化的限制性措施，严格约束开发活动在资源环境承载能力范围内。各省（区、市）应组织开展市、县域资源环境承载能力现状评价，超过承载能力的地区要调整发展规划和产业结构。

实施生态文明绩效评价考核。贯彻落实生态文明建设目标评价考核办法，建立体现生态文明要求的目标体系、考核办法、奖惩机制，把资源消耗、环境损害、生态效益纳入地方各级政府经济社会发展评价体系，对不同区域主体功能定位实行差异化绩效评价考核。

开展环境保护督察。推动地方落实生态环保主体责任，开展环境保护督察，重点检查环境质量呈现恶化趋势的区域流域及整治情况，重点督察地方党委和政府及其有关部门环保不作为、乱作为的情况，重点了解地方落实环境保护党政同责、一岗双责以及严格责任追究等情况，推动地方生态文明建设和环境保护工作，促进绿色发展。

建立生态环境损害责任终身追究制。建立重大决策终身责任追究及责任倒查机制，对在生态环境和资源方面造成严重破坏负有责任的干部不得提拔使用或者转任重要职务，对构成犯罪的依法追究刑事责任。实行领导干部自然资源资产离任审计，对领导干部离任后出现重大生态环境损害并认定其应承担责任的，实行终身追责。

第四节 加强企业监管

建立覆盖所有固定污染源的企业排放许可制度。全面推行排污许可，以改善环境质量、防范环境风险为目标，将污染物排放种类、浓度、总量、排放去向等纳入许可证管理范围，企业按排污许可证规定生产、排污。完善污染治理责任体系，环境保护部门对照排污许可证要求对企业排污行为实施监管执法。2017年底前，完成重点行业及产能过剩行业企业许可证核发，建成全国排污许可管理信息平台。到2020年，全国基本完成排污许可管理名录规定行业企业的许可证核发。

激励和约束企业主动落实环保责任。建立企业环境信用评价和违法排污黑名单制度，企业环境违法信息将记入社会诚信档案，向社会公开。建立上市公司环保信息强制性披露机制，对未尽披露义务的上市公司依法予以处罚。实施能效和环保"领跑者"制度，采取财税优惠、荣誉表彰等措施激励企业实现更高标准的环保目标。到2020年，分级建立企业环境信用评价体系，将企业环境信用信息纳入全国信用信息共享平台，建立守信激励与失信惩戒机制。

建立健全生态环境损害评估和赔偿制度。推进生态环境损害鉴定评估规范化管理，完善鉴定评估技术方法。2017年底前，完成生态环境损害赔偿制度改革试点；自2018年起，在全国试行生态环境损害赔偿制度；到2020年，力争在全国范围内初步建立生态环境损害赔偿制度。

第五节　实施全民行动

提高全社会生态环境保护意识。加大生态环境保护宣传教育，组织环保公益活动，开发生态文化产品，全面提升全社会生态环境保护意识。地方各级人民政府、教育主管部门和新闻媒体要依法履行环境保护宣传教育责任，把环境保护和生态文明建设作为践行社会主义核心价值观的重要内容，实施全民环境保护宣传教育行动计划。引导抵制和谴责过度消费、奢侈消费、浪费资源能源等行为，倡导勤俭节约、绿色低碳的社会风尚。鼓励生态文化作品创作，丰富环境保护宣传产品，开展环境保护公益宣传活动。建设国家生态环境教育平台，引导公众践行绿色简约生活和低碳休闲模式。小学、中学、高等学校、职业学校、培训机构等要将生态文明教育纳入教学内容。

推动绿色消费。强化绿色消费意识，提高公众环境行为自律意识，加快衣食住行向绿色消费转变。实施全民节能行动计划，实行居民水、电、气阶梯价格制度，推广节水、节能用品和绿色环保家具、建材等。实施绿色建筑行动计划，完善绿色建筑标准及认证体系，扩大强制执行范围，京津冀地区城镇新建建筑中绿色建筑达到50%以上。强化政府绿色采购制度，制定绿色产品采购目录，倡导非政府机构、企业实行绿色采购。鼓励绿色出行，改善步行、自行车出行条件，完善城市公共交通服务体系。到2020年，城区常住人口300万以上城市建成区公共交通占机动化出行比例达到60%。

强化信息公开。建立生态环境监测信息统一发布机制。全面推进大气、水、土壤等生态环境信息公开，推进监管部门生态环境信息、排污单位环境信息以及建设项目环境影响评价信息公开。各地要建立统一的信息公开平台，健全反馈机制。建立健全环境保护新闻发言人制度。

加强社会监督。建立公众参与环境管理决策的有效渠道和合理机制，鼓励公众对政府环保工作、企业排污行为进行监督。在建设项目立项、实施、后评价等环节，建立沟通协商平台，听取公众意见和建议，保障公众环境知情权、参与权、监督权和表达权。引导新闻媒体，加强舆论监督，充分利用"12369"环保热线和环保微信举报平台。研究推进环境典型案例指导示范制度，推动司法机关强化公民环境诉权的保障，细化环境公益诉讼的法律程序，加强对环境公益诉讼的技术支持，完善环境公益诉讼制度。

第六节　提升治理能力

加强生态环境监测网络建设。统一规划、优化环境质量监测点位，建设涵盖大气、水、土壤、噪声、辐射等要素，布局合理、功能完善的全国环境质量监测网络，实现生态环境监测信息集成共享。大气、地表水环境质量监测点位总体覆盖80%左右的区县，人口密集的区县实现全覆盖，土壤环境质量监测点位实现全覆盖。提高大气环境质量预报和污染预警水平，强化污染源追踪与解析，地级及以上城市开展大气环境质量预报。建设国家水质监测预警平台。加强饮用水水源和土壤中持久性、生物富集性以及对人体健康危害大的污染物监测。加强重点流域城镇集中式饮用水水源水质、水体放射性监测和预警。建立天地一体化的生态遥感监测系统，实现环境卫星组网运行，加强无人机遥感监测和地面生态监测。构建生物多样性观测网络。

专栏6　全国生态环境监测网络建设

（一）稳步推进环境质量监测事权上收。

对1436个城市大气环境质量自动监测站、96个区域站和16个背景站，2767个国控地表水监测断面、419个近岸海域水环境质量监测点和300个水质自动监测站，40000个土壤环境国家监控点位，承担管理职责，保障运行经费，采取第三方监测服务、委托地方运维管理、直接监测等方式运行，推动环境监测数据联网共享与统一发布。

（二）加快建设生态监测网络。

建立天地一体化的生态遥感监测系统，建立生态功能地面监测站点，加强无人机遥感监测，对重要生态系统服务功能开展统一监测、统一信息公布。建设全国生态保护红线监管平台，建立一批相对固定的生态保护红线监管地面核查点。建立生物多样性观测网络体系，开展重要生态系统和生物类群的常态化监测与观测。新建大气辐射自动监测站400个、土壤辐射监测点163个、饮用水水源地辐射监测点330个。建设森林监测站228个、湿地监测站85个、荒漠监测站108个、生物多样性监测站300个。

加强环境监管执法能力建设。实现环境监管网格化管理，优化配置监管力量，推动环境监管服务向农村地区延伸。完善环境监管执法人员选拔、培训、考核等制度，充实一线执法队伍，保障执法装备，加强现场执法取证能力，加强环境监管执法队伍职业化建设。实施全国环保系统人才双向交流计划，加强中西部地区环境监管执法队伍建设。到2020年，基本实现各级环境监管执法人员资格培训及持证上岗全覆盖，全国县级环境执法机构装备基本满足需求。

加强生态环保信息系统建设。组织开展第二次全国污染源普查，建立完善全国污染源基本单位名录。加强环境统计能力，将小微企业纳入环境统计范围，梳理污染物排放数据，逐步实现各套数据的整合和归真。建立典型生态区基础数据库和信息管理系统。建设和完善全国统一、覆盖全面的实时在线环境监测监控系统。加快生态环境大数据平台建设，实现生态环境质量、污染源排放、环境执法、环评管理、自然生态、核与辐射等数据整合集成、动态更新，建立信息公开和共享平台，启动生态环境大数据建设试点。提高智慧环境管理技术水平，重点提升环境污染治理工艺自动化、智能化技术水平，建立环保数据共享与产品服务业务体系。

专栏7 加强生态环境基础调查

加大基础调查力度，重点开展第二次全国污染源普查、全国危险废物普查、集中式饮用水水源环境保护状况调查、农村集中式饮用水水源环境保护状况调查、地下水污染调查、土壤污染状况详查、环境激素类化学品调查、生物多样性综合调查、外来入侵物种调查、重点区域河流湖泊底泥调查、国家级自然保护区资源环境本底调查、公民生活方式绿色化实践调查。开展全国生态状况变化（2011—2015年）调查评估、生态风险调查评估、地下水基础环境状况调查评估、公众生态文明意识调查评估、长江流域生态健康调查评估、环境健康调查、监测和风险评估等。

第九章
实施一批国家生态环境保护重大工程

"十三五"期间,国家组织实施工业污染源全面达标排放等25项重点工程,建立重大项目库,强化项目绩效管理。项目投入以企业和地方政府为主,中央财政予以适当支持。

专栏8　环境治理保护重点工程

（一）工业污染源全面达标排放。

限期改造50万蒸吨燃煤锅炉、工业园区污水处理设施。全国地级及以上城市建成区基本淘汰10蒸吨以下燃煤锅炉,完成燃煤锅炉脱硫脱硝除尘改造、钢铁行业烧结机脱硫改造、水泥行业脱硝改造。对钢铁、水泥、平板玻璃、造纸、印染、氮肥、制糖等行业中不能稳定达标的企业逐一进行改造。限期改造工业园区污水处理设施。

（二）大气污染重点区域气化。

建设完善京津冀、长三角、珠三角和东北地区天然气输送管道、城市燃气管网、天然气储气库、城市调峰站储气罐等基础设施，推进重点城市"煤改气"工程，替代燃煤锅炉18.9万蒸吨。

（三）燃煤电厂超低排放改造。

完成4.2亿千瓦机组超低排放改造任务，实施1.1亿千瓦机组达标改造，限期淘汰2000万千瓦落后产能和不符合相关强制性标准要求的机组。

（四）挥发性有机物综合整治。

开展石化企业挥发性有机物治理，实施有机化工园区、医药化工园区及煤化工基地挥发性有机物综合整治，推进加油站、油罐车、储油库油气回收及综合治理。推动工业涂装和包装印刷行业挥发性有机物综合整治。

（五）良好水体及地下水环境保护。

对江河源头及378个水质达到或优于III类的江河湖库实施严格保护。实施重要江河湖库入河排污口整治工程。完成重要饮用水水源地达标建设，推进备用水源建设、水源涵养和生态修复，探索建设生物缓冲带。加强地下水保护，对报废矿井、钻井、取水井实施封井回填，开展京津冀晋等区域地下水修复试点。

（六）重点流域海域水环境治理。

针对七大流域及近岸海域水环境突出问题，以580个优先控制单元为重点，推进流域水环境保护与综合治理，统筹点源、面源污染防治和河湖生态修复，分类施策，实施流域水环境综合治理工程，加大整治力度，切实改善重点流域海域水环境质量。实施太湖、洞庭湖、滇池、巢湖、鄱阳湖、白洋淀、乌梁素海、呼伦湖、艾比湖等重点湖库水污染综合治理。开展长江中下游、珠三角等河湖内源治理。

（七）城镇生活污水处理设施全覆盖。

以城市黑臭水体整治和343个水质需改善控制单元为重点，强化污水收集处理与重污染水体治理。加强城市、县城和重点镇污水处理设施建设，加快收集管网建设，对污水处理厂升级改造，全面达到一级A排放标准。推进再生水回用，强化污泥处理处置，提升污泥无害化处理能力。

（八）农村环境综合整治。

实施农村生活垃圾治理专项行动，推进13万个行政村环境综合整治，实施农业废弃物资源化利用示范工程，建设污水垃圾收集处理利用设施，梯次推进农村生活污水治理，实现90%的行政村生活垃圾得到治理。实施畜禽养殖废弃物污染治理与资源化利用，开展畜禽规模养殖场（小区）污染综合治理，实现75%以上的畜禽养殖场（小区）配套建设固体废物和污水贮存处理设施。

（九）土壤环境治理。

组织开展土壤污染详查，开发土壤环境质量风险识别系统。完成100个农用地和100个建设用地污染治理试点。建设6个土壤污染综合防治先行区。开展1000万亩受污染耕地治理修复和4000万亩受污染耕地风险管控。组织开展化工企业搬迁后污染状况详查，制定综合整治方案，开展治理与修复工程示范，对暂不开发利用的高风险污染地块实施风险管控。全面整治历史遗留尾矿库。实施高风险历史遗留重金属污染地块、河道、废渣污染修复治理工程，完成31块历史遗留无主铬渣污染地块治理修复。

（十）重点领域环境风险防范。

开展生活垃圾焚烧飞灰处理处置，建成区域性废铅蓄电池、废锂电池回收网络。加强有毒有害化学品环境和健康风险评估能力建设，建立化学品危害特性基础数据库，建设国家化学品计算毒理中心和国家化学品测试实验室。建设50个针对大型化工园区、集中饮用水水源地等不同类型风险区域的全过程环境风险管理示范区。建设1个国家环境应急救援实训基地，具备人员实训、物资储备、成果展示、应急救援、后勤保障、科技研发等核心功能，配套建设环境应急演练系统、环境应急模拟训练场以及网络培训平台。建设国家生态环境大数据平台，研制发射系列化的大气环境监测卫星和环境卫星后续星并组网运行。建设全国及重点区域大气环境质量预报预警平台、国家水质监测预警平台、国家生态保护监控平台。加强中西部地区市县两级、东部欠发达地区县级执法机构的调查取证仪器设备配置。

（十一）核与辐射安全保障能力提升。

建成核与辐射安全监管技术研发基地，加快建设早期核设施退役及历史遗留放射性废物处理处置工程，建设5座中低放射性废物处置场和1个高放射性废物处理地下实验室，建设高风险放射源实时监控系统，废旧放射源100%安全收贮。加强国家核事故应急救援队伍建设。

专栏9　山水林田湖生态工程

（一）国家生态安全屏障保护修复。

推进青藏高原、黄土高原、云贵高原、秦巴山脉、祁连山脉、大小兴安岭和长白山、南岭山地地区、京津冀水源涵养区、内蒙古高原、河西走廊、塔里木河流域、滇桂黔喀斯特地区等关系国家生态安全的核心地区生态修复治理。

（二）国土绿化行动。

开展大规模植树增绿活动，集中连片建设森林，加强"三北"、沿海、长江和珠江流域等防护林体系建设，加快建设储备林及用材林基地建设，推进退化防护林修复，建设绿色生态保护空间和连接各生态空间的生态廊道。开展农田防护林建设，开展太行山绿化，开展盐碱地、干热河谷造林试点示范，开展山体生态修复。

（三）国土综合整治。

开展重点流域、海岸带和海岛综合整治，加强矿产资源开发集中地区地质环境治理和生态修复。推进损毁土地、工矿废弃地复垦，修复受自然灾害、大型建设项目破坏的山体、矿山废弃地。加大京杭大运河、黄河明清故道沿线综合治理力度。推进边疆地区国土综合开发、防护和整治。

（四）天然林资源保护。

将天然林和可以培育成为天然林的未成林封育地、疏林地、灌木林地全部划入天然林，对难以自然更新的林地通过人工造林恢复森林植被。

（五）新一轮退耕还林还草和退牧还草。

实施具备条件的25度以上坡耕地、严重沙化耕地和重要水源地15—25度坡耕地退耕还林还草。稳定扩大退牧还草范围，优化建设内容，适当提高中央投资补助标准。实施草原围栏1000万公顷、退化草原改良267万公顷，建设人工饲草地33万公顷、舍饲棚圈（储草棚、青贮窖）30万户、开展岩溶地区草地治理33万公顷、黑土滩治理7万公顷、毒害草治理12万公顷。

（六）防沙治沙和水土流失综合治理。

实施北方防沙带、黄土高原区、东北黑土区、西南岩溶区以及"一带一路"沿线区域等重点区域水土流失综合防治,以及京津风沙源和石漠化综合治理,推进沙化土地封禁保护、坡耕地综合治理、侵蚀沟整治和生态清洁小流域建设。新增水土流失治理面积27万平方公里。

(七)河湖与湿地保护恢复。

加强长江中上游、黄河沿线及贵州草海等自然湿地保护,对功能降低、生物多样性减少的湿地进行综合治理,开展湿地可持续利用示范。加强珍稀濒危水生生物、重要水产种质资源以及产卵场、索饵场、越冬场、洄游通道等重要渔业水域保护。推进京津冀"六河五湖"、湖北"四湖"、钱塘江上游、草海、梁子湖、汾河、滹沱河、红碱淖等重要河湖和湿地生态保护与修复,推进城市河湖生态化治理。

(八)濒危野生动植物抢救性保护。

保护和改善大熊猫、朱鹮、虎、豹、亚洲象、兰科植物、苏铁类、野生稻等珍稀濒危野生动植物栖息地,建设原生境保护区、救护繁育中心和基因库,开展拯救繁育和野化放归。加强野外生存繁衍困难的极小种群、野生植物和极度濒危野生动物拯救。开展珍稀濒危野生动植物种质资源调查、抢救性收集和保存,建设种质资源库(圃)。

(九)生物多样性保护。

开展生物多样性保护优先区域生物多样性调查和评估,建设50个生物多样性综合观测站和800个观测样区,建立生物多样性数据库及生物多样性评估预警平台、生物物种查验鉴定平台,完成国家级自然保护区勘界确权,60%以上国家级自然保护区达到规范化建设要求,加强生态廊道建设,有步骤地实施自然保护区核心区、缓冲区生态移民，完善迁地保护体系，建设国家生物多样性博物馆。开展生物多样性保护、恢复与减贫示范。

（十）外来入侵物种防治行动。

选择50个国家级自然保护区开展典型外来入侵物种防治行动。选择云南、广西和东南沿海省份等外来入侵物种危害严重区域，建立50个外来入侵物种防控和资源化利用示范推广区，建设100个天敌繁育基地、1000公里隔离带。建设300个口岸物种查验点，提升50个重点进境口岸的防范外来物种入侵能力。针对已入侵我国的外来物种进行调查，建立外来入侵物种数据库，构建卫星遥感与地面监测相结合的外来入侵物种监测预警体系。

（十一）森林质量精准提升。

加快推进混交林培育、森林抚育、退化林修复、公益林管护和林木良种培育。精准提升大江大河源头、国有林区（场）和集体林区森林质量。森林抚育4000万公顷，退化林修复900万公顷。

（十二）古树名木保护。

严格保护古树名木树冠覆盖区域、根系分布区域，科学设置标牌和保护围栏，对衰弱、濒危古树名木采取促进生长、增强树势措施，抢救古树名木60万株、复壮300万株。

（十三）城市生态修复和生态产品供给。

对城市规划区范围内自然资源和生态空间进行调查评估，综合识别已被破坏、自我恢复能力差、亟需实施修复的区域，开展城市生态修复试点示范。推进绿道绿廊建设，合理规划建设各类公园绿地，加快老旧公园改造，增加生态产品供给。

（十四）生态环境技术创新。

建设一批生态环境科技创新平台，优先推动建设一批专业化环保高新技术开发区。推进水、大气、土壤、生态、风险、智慧环保等重大研究专项，实施京津冀、长江经济带、"一带一路"、东北老工业基地、湘江流域等区域环境质量提升创新工程，实施青藏高原、黄土高原、北方风沙带、西南岩溶区等生态屏障区保护修复创新工程，实施城市废物安全处置与循环利用创新工程、环境风险治理与清洁替代创新工程、智慧环境创新工程。推进环境保护重点实验室、工程技术中心、科学观测站和决策支撑体系建设。建设澜沧江—湄公河水资源合作中心和环境合作中心、"一带一路"信息共享与决策平台。

第十章
健全规划实施保障措施

第一节 明确任务分工

明确地方目标责任。地方各级人民政府是规划实施的责任主体,要把生态环境保护目标、任务、措施和重点工程纳入本地区国民经济和社会发展规划,制定并公布生态环境保护重点任务和年度目标。各地区对规划实施情况进行信息公开,推动全社会参与和监督,确保各项任务全面完成。

部门协同推进规划任务。有关部门要各负其责,密切配合,完善体制机制,加大资金投入,加大规划实施力度。在大气、水、土壤、重金属、生物多样性等领域建立协作机制,定期研究解决重大问题。环境保护部每年向国务院报告环境保护重点工作进展情况。

第二节 加大投入力度

加大财政资金投入。按照中央与地方事权和支出责任划分的要求,加快建立与环保支出责任相适应的财政管理制度,各级财政应保障同级生态环保重点支出。优化创新环保专项资金使用方式，加大对环境污染第三方治理、政府和社会资本合作模式的支持力度。按照山水林田湖系统治理的要求，整合生态保护修复相关资金。

拓宽资金筹措渠道。完善使用者付费制度，支持经营类环境保护项目。积极推行政府和社会资本合作，探索以资源开发项目、资源综合利用等收益弥补污染防治项目投入和社会资本回报，吸引社会资本参与准公益性和公益性环境保护项目。鼓励社会资本以市场化方式设立环境保护基金。鼓励创业投资企业、股权投资企业和社会捐赠资金增加生态环保投入。

第三节 加强国际合作

参与国际环境治理。积极参与全球环境治理规则构建，深度参与环境国际公约、核安全国际公约和与环境相关的国际贸易投资协定谈判，承担并履行好同发展中大国相适应的国际责任，并做好履约工作。依法规范境外环保组织在华活动。加大宣传力度，对外讲好中国环保故事。根据对外援助统一部署，加大对外援助力度，创新对外援助方式。

提升国际合作水平。建立完善与相关国家、国际组织、研究机构、民间团体的交流合作机制，搭建对话交流平台，促进生态环保理念、管理制度政策、环保产业技术等方面的国际交流合作，全面提升国际化水平。组织开展一批大气、水、土壤、生物多样性等领域的国际合作项目。落实联合国2030年可持续发展议程。加强与世界各国、区域和国际组织在生态环保和核安全领域的对话交流与务实合作。加强南南合作，积极开展生态环保和核安全领域的对外合作。严厉打击化学品非法贸易、固体废物非法越境转移。

第四节 推进试点示范

推进国家生态文明试验区建设。以改善生态环境质量、推动绿色发展为目标,以体制创新、制度供给、模式探索为重点,设立统一规范的国家生态文明试验区。积极推进绿色社区、绿色学校、生态工业园区等"绿色细胞"工程。到2017年,试验区重点改革任务取得重要进展,形成若干可操作、有效管用的生态文明制度成果;到2020年,试验区率先建成较为完善的生态文明制度体系,形成一批可在全国复制推广的重大制度成果。

强化示范引领。深入开展生态文明建设示范区创建,提高创建规范化和制度化水平,注重创建的区域平衡性。加强创建与环保重点工作的协调联动,强化后续监督与管理,开展成效评估和经验总结,宣传推广现有的可复制、可借鉴的创建模式。

深入推进重点政策制度试点示范。开展农村环境保护体制机制综合改革与创新试点。试点划分环境质量达标控制区和未达标控制区,分别按照排放标准和质量约束实施污染源监管和排污许可。推进环境审计、环境损害赔偿、环境服务业和政府购买服务改革试点,强化政策支撑和监管,适时扩大环境污染第三方治理试点地区、行业范围。开展省级生态环境保护综合改革试点。

第五节 严格评估考核

环境保护部要会同有关部门定期对各省(区、市)环境质量改善、重点污染物排放、生态环境保护重大工程进展情况进行调度，结果向社会公开。整合各类生态环境评估考核，在 2018 年、2020 年底，分别对本规划执行情况进行中期评估和终期考核，评估考核结果向国务院报告，向社会公布，并作为对领导班子和领导干部综合考核评价的重要依据。

The 13th Five-Year Plan for the Protection of Ecological Environment

Circular of the State Council on Printing and Distributing the 13th Five-Year Plan for the Protection of Ecological Environment

SC [2016] No. 65

To the people's governments of all provinces, autonomous regions and municipalities directly under the Central Government, ministries and commissions of the State Council and departments directly under the State Council,

The 13th Five-Year Plan for the Protection of Ecological Environment is hereby printed and distributed to you for thorough implementation.

The State Council

November 24, 2016

(This is a revised version of the original.)

(*This is not an official translation of the Chinese original and is published for reader's reference only.*)

Chapter 1 An Overview of Environmental Protection in China

The Central Committee of the Communist Party (CPC) and the State Council of China attach great importance to environmental protection. As the "War on Pollution" unfolded at the beginning of the 12th Five-Year Plan (FYP) period, the authorities have taken a resolute stance against air, water and soil contamination and doubled the efforts to protect the ecological environment and improve environmental quality. Thanks to these efforts, environmental quality in China has seen significant improvement and all major targets and tasks set forth in the 12th Five-Year Plan for Environmental Protection have been accomplished. Yet despite of all these progress, in the 13th FYP period, China still faces serious challenges of unbalanced, uncoordinated and unsustainable economic and social development, intertwined with complex environmental problems that are rooted in different stages of development but occurring all at the same time in a spectrum of sectors and of various types. Given the considerable gap between current environmental quality and people's needs and expectations, it remains China's priority to improve environmental quality and governance capacity to overcome the weak link of ecological environment.

Section 1 Positive progress

Ecological civilization as a national strategy. The CPC Central Committee and the State Council give high priority to promoting ecological civilization. President Xi Jinping stressed on many occasions that "Green is Gold", "we should adhere to the basic state policy of conserving resources and protecting the environment", and that "we should protect the ecological environment like protecting our eyes, and treat the environment like it is our lives". Premier Li Keqiang repeatedly called for efforts to strengthen integrated environmental governance, promote ecological civilization and advance green development, and reiterated the government's determination to set on a win-win path for both economic development and environmental protection. Since the 18th CPC National Congress, the CPC Central Committee and the State Council have placed the development of ecological civilization high on its national agenda, and incorporated it into the overall plan for promoting economic, political, cultural, social, and ecological progress, followed by a range of policies and plans from promulgating the *Integrated Reform Plan for Promoting Ecological Progress,* to implementing action plans for prevention and control of air, water and soil pollution. By that, the vision of development, of governance and the understanding of nature have been made highly consistent with each other, and by integrating all these into the overall governance and development philosophies, China saw a great leap forward in the understanding, practice, and progress of ecological civilization.

Moderate improvement in environmental quality. Air pollution control and prevention begin to bear fruit. In 2015, the annual average concentration of fine particulate matter ($PM_{2.5}$) of 338 cities at the prefecture level or above registered 50 $\mu g/m^3$. Compared with 2013, the annual average $PM_{2.5}$ concentration of 74 monitored cities went down by 23.6% and the levels of the Beijing-Tianjin-Hebei region, the Yangtze River Delta (YRD) and the Pearl River Delta (PRD) went down by 27.4%, 20.9% and 27.7% respectively. The proportion of acid rain area fell to 7.6% from a historical peak of 30%. There is also a significant improvement of water quality of main streams and rivers, with 66% of 1,940 national monitoring sections meeting Grade I-III Standards and only 9.7% worse than Grade V. The forest coverage rate has reached 21.66% with forest stock volume at 15.14 billion m^3, and the vegetation coverage rate of grassland increased to 54%. A total of 2,740 nature reserves, taking up 14.8% of the land area, have been built, providing protection for over 90% of terrestrial natural ecosystems, 89% of national key protected wild animals and plant species as well as for a majority important natural monuments. The population of rare and endangered species such as giant panda, Manchurian tigers, crested ibis, Tibetan antelope and Yangtze alligator remain stable with some increase. There has been a reduction of both desertificated and sandificated land areas for three consecutive monitoring cycles.

Pollution control and emission reduction targets overachieved. In 2015, 99% and 92% of the total installed capacity of the coal-fired power generators were equipped with desulfurization and denitrification respectively, and coal-fired power generators with an installed capacity of 160 million kW transformed for ultra-low emissions. The urban sewage treatment rate was increased to 92% and the rate of solid waste subject to environmentally sound disposal reached 94.1% in built-up urban areas. A total of 72,000 villages have taken measures to comprehensively improve environmental, directly benefiting more than 120 million rural residents. A total of 61,000 large-scale breeding farms (districts) have installed waste treatment and recycling facilities. There was 12.9% reduction of COD discharge, 13% reduction of ammonia nitrogen, 18% reduction of sulfur dioxide (SO_2) and 18.6% reduction of nitrogen oxide (NO_x) in accumulative terms during the 12th FYP period.

Positive progress in conserving and enhancing ecosystems. A number of major ecological conservation and restoration projects have been steadily carried out, including projects on natural forests protection, "grain for green", shelter forests, protection and restoration of rivers, lakes and wetlands, soil and water conservation, desertification and stony desertification control, wildlife protection and development of nature reserves. Commercial logging of natural forests has been ceased in all key state-owned forest areas. The protected area of wetlands has gone up by 5.2594 million ha. and the protection rate of natural wetland protection increased to 46.8%. Effort has also been made on improving 100,000 km^2 of desertificated land and 266,000 km^2 of soil erosion. Report on *National Ecosystem Survey and Assessment of China (2000-2010)* have been finished, and *China Biodiversity Red List* released. More than 4,300 forest parks, wetland parks and desert parks at different levels have been established. A total of 16 provinces, autonomous regions and municipalities have made their efforts towards Ecological Province and more than 1,000 cities, counties and urban districts have taken measures to develop Ecological Cities Counties or Urban Districts, with 114 cities counties and districts winning the title of National Demonstration Sites on Ecological Development. The government has released and circulated a plan for reforming state-owned forest farms and the guidelines for reforming state-owned forest areas, and six provinces have successfully concluded their trial reforms on state-owned forest farms.

Environmental risks under effective control. As of 2015, a total of 50 facilities for hazardous waste disposal and 273 facilities for collective disposal of medical waste have been built and concluded, with all 6.7 million tons of chromium slag left from past production successfully disposed. The discharge of lead, mercury, cadmium, chromium and arsenic went down by 27.7% compared with that of 2007, greatly bringing down the number of heavy metal pollution emergency accidents. The environmental impact of Tianjin Port Explosions on August 12 in 2015 was minimized through a science-based approach. Nuclear facilities were made safer, and nuclear technologies were used and managed in full compliance with relevant standards and procedures. Radiation environmental quality remains good.

A sound legal framework on ecological environment. China has promulgated and amended major environmental laws including the *Environmental Protection Law*, the *Law on Prevention and Control of Air Pollution*, *Regulations on Safe Management of Radioactive Waste* and *Air Quality Standard*. Documents such as the *Measures on Accountability for Damage of Ecological Environment* have been released and the ecological compensation mechanism further improved. There are continued efforts to implement activities planned for the implementation year of new *Environmental Protection Law*, and to conduct environmental protection inspection. The entire society is more aware of the significance of rule of law on ecological environment.

Section 2 Ecological environment remaining a weak link in achieving a moderately prosperous society in all aspects

Heavy and spreading pollutants discharge coupled with severe environmental pollution. The COD and SO_2 emissions stay at a high level of around 20 million tons, and the environmental carrying capacity exceeds or approaches the upper limit. Up to 78.4% of cities fail to meet air quality standards, and 3.2% of the days are under severe pollution or worse that sparked great public concerns. There is frequent heavy air pollution in winter in some areas. Urgent steps need to be taken to protect drinking water sources. Sewage outfalls are not properly distributed in a way commensurate with the carrying capacity of water environment, leaving black and putrid waters in urban built-up areas. There is still prominent eutrophication of lakes and reservoirs and heavy water pollution in some river basins. 16.1% of monitoring sites of soil and 19.4% of monitoring sites of farmland fail to meet national soil quality standards. Soil contamination is severe in solid waste sites and abandoned mining sites. There is a wide gap in urban-rural environmental public services, adding to the difficulty in managing and improving the environment.

Poor coordination in protecting mountains, waters, forests and farmlands causing huge ecological damage. Up to 55% of national land is ecologically venerable at a moderate level or above, and nearly 20% of land suffers from desertification or stony desertification. Forest systems are degrading with a poor forest quality, pure forests still dominate, and the ecosystems fail to provide efficient service. It's seen in China a growing number of artificial landscape. About 1,333 km^2 of forest land is illegally occupied every year and the forest stock volume per unit of forest area is only 78% of the global average. Grassland degradation has not been fundamentally reversed. More than 1/3 of grasslands are under intermediate and heavy degradation, while the restored prairie ecosystems remain vulnerable. The total wetland areas decrease by about 5.1 million mu (1 mu is equivalent to 0.0667 hectares) annually in recent years. More than 900 species of vertebrates and 3,700 species of higher plants are threatened. Resources are overused and exploited, which greatly impairs the ecosystems, encroaches the ecological space on a continuous basis, and damages the ecological resources in some areas, thus making the protection efforts even more challenging.

Poor industrial structure and layout posing high environmental risk. As a major chemical producer and consumer, with toxic and hazardous pollutants becoming increasingly diversified, China is facing ever growing environmental risks that are manifested at regional, structural and spatial planning level. There are frequent environmental pollution incidents as a result of hazardous chemicals accidents, and a large number of environment-risky enterprises located near to waters and cities are posing great threat to the environment. The reasons for environmental emergencies are increasingly complex and the pollutants involved are diversified. They are impacting larger scope of areas, among which more are environmentally sensitive. Environmental pollution accidents are increasingly characterized as complexity, diversified pollutants and widespread impacts that are more often taking place in ecologically sensitive areas. In the last decade, there were over 7,600 forest fires annually and 175 million mu of forests hit by pests and insects. In recent years, an annual average of one million batches of harmful pests have been intercepted, a number that indicates high risks of animal and plant infection and of the invasion of quarantine pests from national border.

Section 3 Opportunities and challenges for the protection of ecological environment

China will embrace an important strategic opportunity for protecting the ecological environment in the 13th FYP period. It is expected that dividends from policies, laws and technologies for environmental protection will be unleashed with efforts to advance comprehensive reforms, law-based governance, innovation- and green- oriented development, and the systems and mechanisms for ecological civilization. With the economy restructured and upgraded, and supply-side structural reform gathering pace, China will move fast in cutting excess capacity of heavy pollution, expanding supply of eco-products and easing the pressure brought about by incremental pollutants discharge. With a growing public awareness of environmental protection, all parts of the society are now pulling together for a sound environment for all.

However, as industrialization, urbanization and agricultural modernization are still underway, it remains challenging to protect the environment at the same time. With mounting downward pressure, the existing conflict between development and environment becomes even more acute. Some localities have to cut its environmental expenditure, imposing more pressure on environmental governance and improving environmental quality. There is a growing regional disparities in terms of ecological environment, and pollution is spreading from one single source in a small area to one that covers a wider range of different areas. Plus that ecosystem in some localities are becoming less stable and less capable of providing ecological services, it's rather difficult to coordinate protection efforts in different areas. With China actively combating global climate change and pushing forward the "Belt and Road Initiative (BRI)", the international community especially the developed countries are calling on China to take on more environmental responsibilities, so China is faced with ever-growing challenges in fully engaging in global environmental governance.

In short, the 13th FYP period will present both opportunities and challenges for China. It's a crucial period when China has to press ahead and is very likely to make significant progress under great pressure, as well as an opportune time to make breakthroughs in improving environmental quality. China should make full use of these new opportunities and favorable conditions to properly address various risks and challenges, and advance environmental protection efforts to improve the overall environmental quality.

Chapter 2 Guidelines, Basic Principles and Main Objectives

Section 1 Guidelines

The guidelines of the *Plan* are providing people with more high-quality ecological products and contributing to the realization of the Two Centenary Goals and the Chinese Dream of great rejuvenation of the Chinese nation through thoroughly implementing the guiding principles from the 18th National Congress of the Communist Party of China (CPC) and those of the third through sixth plenary sessions of the 18th Party Central Committee; following the guidance of *Deng Xiaoping Theory*, the *Theory of the Three Represents*, and the *Scientific Outlook on Development*; putting into practice the guiding principles from the major speeches of General Secretary Xi Jinping and his new vision, thinking, and strategies for China's governance; promoting balanced economic, political, cultural, social, and ecological progress and coordinated implementation of the *Four-Pronged Comprehensive Strategy*; as well as adhering to and promoting the innovative, coordinated, green, open and shared development. In accordance with the decisions and plans of the CPC Central Committee and State Council, the core is to improve the overall environmental quality, which calls for the most stringent environmental protection system and intensified efforts to combat air, water and soil pollution. It also requires steps to protect and restore ecosystems, strictly control and guard against ecological risks, as well as modernize national environmental governance system and strengthen its capacity by implementing a management system that focuses on systematic approach and rule of law, and is science-based and detail-oriented, as well as supported by information technologies.

Section 2 Basic principles

Upholding green development as a fundamental approach to various environmental issues. Green development makes a country prosper and delivers tangible benefits to its people. It calls for balanced relations between development and environmental protection, and for coordination with the development of a new type of industrialization, urbanization, information technology, and agricultural modernization. Based on the immediate needs while looking ahead to the future, environmental protection must be incorporated into the efforts to maintain steady economic growth, adjust economic structure, improve people's livelihood and prevent risks. China will strengthen control at source through advancing supply-side structural reform, optimizing spatial layout, and promoting green way of life and production, so that ecological damage and environmental pollution are able to be prevented at source. There will also be moves to strengthen governance over ecological environment to promote harmonious development between man and nature.

Taking systematic approaches with quality improvement at its core. Aiming at addressing prominent environmental issues, it's necessary to clearly define the objectives and tasks of improving environmental quality for each region, each watershed and at each different stage. It's encouraged to adopt multiple measures of structural adjustment, pollution control, emission reduction, up-to-standard discharge and ecological conservation. There will be a number of major projects in place, such as integrated planning on preventing and controlling multiple pollutants, and restoring ecosystems and tackling environmental problems in a systematic way, in a view to steadily improve the environmental quality and build up capacity in supplying quality ecological products.

Optimizing spatial planning and management based on specific situations. With ecological conservation as the top priority, space for production, life and ecological protection will be properly planned and managed in a holistic manner. This requires steps to set ecological red lines (ERLs) and to hold firm to these lines so as to uphold national ecological security. There will be a full-fledged management system in line with the needs for a detail-oriented management, where rights and responsibilities are explicit, regulation is effective, and differentiated approaches are implemented based on regional disparities and categories and in a tiered and step-by-step manner.

Sticking to reform and innovation while enhancing rule of law. Reform and innovation must serve as a driving force of environmental protection to enable a shift of mindset and model of environmental governance. It includes efforts to reform the basic systems of environment governance, establish a business discharge permit system that covers all stationary sources, and push forward a vertical management of environmental protection agencies on monitoring and supervision that are below the provincial level. With all these, it's expected to put in place a full-fledged institutional system on ecological civilization at an early date. China will strengthen environmental legislation, justice and law enforcement, and take tough and forceful measures to ensure public compliance. Laws, regulations and systems will help to protect the environment through preventing pollution at source, regulating the operations, and imposing stringent punishment on law violations.

Performing environmental duties and encouraging multi-governance. China will introduce stringent liability systems on environmental protection, by which the powers and expenditure responsibilities of central and local environmental authorities will be clearly defined, both the party and the government officials would be held responsible for environmental issues, and leaders and officials are responsible for both economic growth and environmental protection. Enterprises will be the major entity responsible for addressing environmental pollution, and the entire society will be mobilized to participate in environmental protection. There will be both carrots and sticks in place, and the role of both the government and the market will be brought into full play. All of these will contribute to a multi-governance system on environmental protection that involves the government, business and civil society.

Section 3 Main objectives

The overall objective is to improve the environmental quality by 2020. This includes specified targets of promoting a green life and production, advancing low-carbon development, notably bringing down total discharge of major pollutants, effectively controlling environmental risks, reversing biodiversity loss, striving for a more stable ecosystem, building ecological-security shields, achieving significant strides in modernizing national environmental governance system and capacity, and of bringing ecological civilization more aligned with the goal of achieving a moderately prosperous society in all aspects.

Box 1. Major Targets of the 13th Five-Year Plan for Protecting the Ecological Environment

Indicators		2015	2020	Cumulative [1]	Nature
Eco-environmental quality					
1. Air quality	Percent of days with good air quality in cities at prefecture-level and above[2] (%)	76.7	> 80	-	Binding
	Reduction of fine particle concentration in cities at prefecture-level or above failing to meet the standard (%)	-	-	(18)	Binding
	Decline in the percent of days with heavy pollution or even worse in cities at prefecture-level or above (%)	-	-	(25)	Anticipated
2. Water quality	Percent of surface water with quality at or better than Grade III[3] (%)	66	> 70	-	Binding
	Percent of surface water with quality worse than Grade V (%)	9.7	< 5	-	Binding
	Percent of major rivers and lakes attaining water quality standards (%)	70.8	> 80		Anticipated

Indicators		2015	2020	Cumulative [1]	Nature
2. Water quality	Percent of groundwater with very poor quality (%)	15.7[4]	Around 15	-	Anticipated
	Percent of coastal waters with excellent and good water quality (I, II) (%)	70.5	Around 70	-	Anticipated
3. Soil quality	Safe utilization rate of contaminated farmland (%)	70.6	Around 90	-	Binding
	Safe utilization rate of contaminated fields (%)	-	Above 90	-	Binding
4. Ecological conditions	Forest coverage (%)	21.66	23.04	(1.38)	Binding
	Forest stock volume (100 million m^3)	151	165	(14)	Binding
	Stock wetland (667 km^2)	-	≥8	-	Anticipated
	Vegetation coverage of grassland (%)	54	56		Anticipated
	Environmental condition index of counties in areas with key ecological functions	60.4	> 60.4	-	Anticipated
Total pollutants discharge					
5. Reduction of major pollutants discharge (%)	Chemical oxygen demand (COD)	-	-	(10)	Binding
	Ammonia nitrogen	-	-	(10)	
	Sulfur dioxide (SO$_2$)	-	-	(15)	
	Nitrogen oxides (NOx)	-	-	(15)	

Indicators		2015	2020	Cumulative [1]	Nature
6. Reduction of regional pollutants discharge (%)	Volatile organic compounds (VOCs) of key industries in key regions [5]	-	-	(10)	Anticipated
	Total nitrogen (TN) in key regions [6]	-	-	(10)	Anticipated
	Total phosphorous (TP) in key regions [7]	-	-	(10)	
Ecological conservation and restoration					
7. Protection rate of wildlife under national priority protection (%)		-	> 95	-	Anticipated
8. Natural shoreline retention rate (%)		-	≥35	-	Anticipated
9. Newly protected land under desertification control (10,000 km^2)		-	-	(10)	Anticipated
10. Newly protected land under water and soil erosion control (10,000 km^2)		-	-	(27)	Anticipated

Note: 1. Five-year cumulative number in brackets ( ).

2. Air quality assessment covers 338 cities nationwide (including prefecture-level, league-level and some county-level cities under provincial jurisdiction, excluding Sansha and Danzhou).

3. Water environmental quality assessment covers surface water sections under national monitoring program which increased from 972 (during the 12th FYP period) to 1,940.

4. Data of 2013.

5. More than 10% of total VOCs emission would be cut through strengthened control over its emission in key industries and key regions.

6. Total TN control covers 56 coastal cities and 29 eutrophic lakes and reservoirs.

7. Total TP control covers units with excessive TP emissions and related upstream areas.

Chapter 3 Prevent and Control Pollution at Source to Lay Groundwork for Green Development

Green development holds the key to breaking the bottleneck of limited natural resources and improving the quality of development. It's therefore crucial to innovate regulation measures, and enhance management at source. To this end, China will properly manage ecological space to optimize spatial layout in line with green development, protect the environment to catalyze supply-side structural reform, and innovate in green technologies to facilitate environmental governance. With all these efforts set in place, it's aimed to promote green and coordinated development in key regions, optimize spatial planning and industrial structure conducive to environmental protection and resource conservation, and develop a green way of life and production to protect the environment from the source.

Section 1 Strengthen management and control of ecological space

Thoroughly implementing function-oriented zoning. Function-oriented zoning shall play a greater role in developing and protecting China's territorial space. It's therefore crucial to develop a spatial layout based on functions. Objectives, measures and assessment criteria regarding environment protection will be developed according to the main functions of different regions. To be specific, for those regions where development is strictly prohibited, mandatory steps must be taken to control human activities from damaging the natural ecosystem and natural cultural heritage, to prohibit all kinds of development activities that work against the region's major function, and to resettle the local population in an orderly and gradual manner. In important ecological areas where development is restricted, there will be efforts to control development intensity, develop an environment-friendly industrial structure, as well as maintain and enhance its capacities in supplying ecological products and providing ecological service. In major grain producing areas where development is restricted, priority shall be given to protecting the soil environment of farmland to ensure the supply, quality and safety of agricultural products. In areas where major development activities are taking place, environmental management and governance will be strengthened, emission intensity of pollutants will be reduced substantially, negative impacts exerted by industrialization and urbanization on the ecological environment mitigated, and living environment and environment quality improved. In areas where development remains a priority, cities will be developed in an intensive, compact, green and low-carbon manner, where green ecological space will be expanded, and ecological layout will be optimized. Function-oriented zoning will also target oceans and seas to better use and protect the marine resources.

Setting and holding firm to redlines for ecological conservation. Redlines for ecological conservation will be set for provinces and municipalities in the Beijing-Tianjin-Hebei region and along the Yangtze River Economic Belt by the end of 2017, and in the rest provinces, autonomous regions and municipalities by the end of 2018. China will complete the survey and delineate lines for ecological conservation across the country, putting in place a redlining system by the end of 2020, and implement measures to control and mange those red lines. There will also be steps to set in place and refine the ecological compensation mechanism, and release information on redlines on a regular basis. A monitoring and performance evaluation systems will be developed to assess performances of each province, autonomous region and municipality. It also calls for measures to uphold national ecological security, including strengthening ecological functions of forests, grasslands, rivers, lakes, wetlands and marine waters, and enhancing their capacity in providing quality ecological products.

Promoting "integrated planning". Based on function-oriented zoning, China will better regulate development activities through introducing procedures and stringent standards in terms of ecological environment space, ecological and environmental carrying capacity, bottom line for environmental quality, and strategic environmental impact assessment (SEA) and environmental impact assessment (EIA) on development plans. Technical specifications for ecological conservation redlines, environmental quality bottom lines, resources utilization ceilings and negative lists of access will be developed and implemented to push forward integrated planning. With each administrative region at municipal level or county-level taken as a unit, a spatial governance system composed of spatial planning, space use control and differentiated performance evaluation will be established. There shall be a national spatial planning system in place to coordinate and integrate various spatial plans in a bid to promote "integrated planning". It's also planned to study and roll out guidelines on promoting "integrated planning" on protecting ecological environment, and starting from 2018, initiate studies on spatial planning for better protecting the ecosystems and environment in provinces, regions and city clusters.

Section 2 Advance supply-side structural reform

Cutting backward and excess capacity through hard-line measures. China will establish a mechanism to retire heavy polluting capacity as well as to cut excess capacity, and shut down in accordance with law those enterprises that have been continuously discharging excessive pollutants, unable and unwilling to address pollutions, and remaining high in pollution despite of sound treatment. There will be continued efforts to revise and refine *The Comprehensive Directory of Environmental Protection*, and phase out those production technologies, equipment and products that generate heavy pollution or pose high environmental risks. Localities are encouraged to develop policies that cover more sectors and impose more stringent standards in phasing out backward production capacity. On the part of Beijing-Tianjin-Hebei region, there will be double efforts to phase out excess capacity in industries like iron and steel that fail to meet discharge standards. Based on resource and environment carrying capacity of different regions, growth in industries of paper making, leather, printing and dyeing, coking, sulfur refining, arsenic refining, oil refining, electroplating and pesticides must subject to the development limits. Newly constructed, renovated or expanded project won't be approved unless its main pollutants discharge is equal to or less than that of phased out ones. It's also required to adjust and optimize industrial structure and replace backward and excessive capacity with projects of equal or lower capacity in industries of coal, steel, cement and flat glass.

Upgrading and transforming enterprises with stringent energy consumption requirements. Both the total energy consumption and the consumption intensity will be controlled to conserve energy in key sectors of manufacturing, construction, transportation and public institutions. Strict evaluation and examination on energy conservation will be conducted on newly-launched projects, and the performance of manufacturing industries will be supervised and regulated throughout its entire life cycle. Targeting traditional manufacturing industries such as iron and steel, non-ferrous metals, chemicals, building materials, light industry and textile, technologies concerning improving energy efficiency of electric engines and transformers, clean production, water saving, pollution treatment and recycling will be upgraded. A group of key energy-conserving projects will be launched, including those on improving energy efficiency of generator systems, comprehensively upgrading coal-fired boilers for energy conservation and environment protection, and promoting green lighting and residual heat utilization. There will be supporting policies to make enterprises more capable of clean and green manufacturing, and encourage industrial parks and enterprises to apply distributed energy systems.

Promoting green manufacturing as well as production and supply of green products. A product's entire life cycle ranging from its design to raw material selection, production, procurement, logistics and recycling must all go green. Enterprises are encouraged to practice green design, develop green products, improve green packaging standard system, and promote minimum and pollution-free packaging and packaging material recycling. Steps will be taken to put in place a green system for manufacturing industry, which includes developing green plants, green industrial parks, green supply chains, green evaluation and green techniques. Meanwhile, it also calls for measures to expand green products supply, and to integrate various certifications of environmentally friendly, energy-efficient, water-saving, recyclable, low-carbon, renewable and organic products to establish a set of unified standards, certification and labeling system for these products. To boost ecological agriculture and organic agriculture, there will be enabling measures to develop organic food bases and industries in a bid to supply more organic products. By all these steps, it's aimed by 2020 there will be a full-fledged green manufacturing system in place, with 100 Model Enterprises on Green Design, 100 Green Industrial Demonstration Parks, and 100 Green Demonstration Plants.

Promoting circular development. China will implement the "Initiative to Guide a Shift towards Circular Development", which includes steps to collectively dispose municipal low-value waste, develop resource-recycling demonstration bases and ecological industrial parks, and launch some industrial demonstration sites on new type of industrialization and circular economy, as well as develop demonstration cities and counties for circular economy. There will also be demonstration projects on high-end, intelligent and in-service remanufacturing, and more pilot projects to comprehensively utilize and recycle the resources, and build bases for such purpose. Based on the progress made through developing national demonstration centers for recovering mineral resources from urban waste, some key enterprises, industrial bases and parks for recycling and comprehensively utilizing renewable resources will be developed. The initiative also calls for better management of recycling industries on waste iron and steel, tires, textiles and clothing, plastics and power batteries to enhance the network for renewable resource recycling and utilization. There will be measures trying to create a reverse-recycling channel, and promote new recycling methods such as "Internet + Recycle" and intelligent recycling. An extended producer responsibility system is expected to be set in place. It's planned by 2020 the comprehensive utilization rate of industrial solid waste of China will reach 73%, and no more fertilizers will be used. There will be demonstration projects on circular agriculture to promote the commercial use of straw and efforts to develop it into an industry. It's also aimed by 2020, 85% of straw will be comprehensively utilized, and basically all agricultural resources recycled in national demonstration sites on modern agriculture and major grain-producing counties.

Promoting energy-conserving and environment-friendly industries. China will facilitate research, development and commercialization of core environmental technologies, complete sets of products, facilities and equipment, materials and reagents that serve to advance low-carbon and circular development, pollution control and emissions reduction, as well as monitoring and supervision. By that, it's expected to develop as soon as possible a number of leading technologies and products that are able to compete on the international market. It's encouraged to develop specialized services such as technical advisory, system design, equipment manufacturing, project implementation, and business management for energy conservation and environmental protection. Environmental services will be promoted through developing service markets relevant to energy performance contracting, water saving performance contracting, third-party monitoring, third-party pollution treatment, and public-private partnership (PPP). There will be pilot projects to introduce the third party into managing the environment of small towns and industrial parks, and regulations over the management of environmental performance contracts as part of the efforts to develop an environmental service performance evaluation mechanism. A government procurement list on environmental service will be developed, and social capital is encouraged to invest in environment-friendly enterprises. There will be moves to develop some large-scale energy-conserving and environment-friendly enterprises and brands that are internationally competitive. China will encourage its people to start business and make innovations in the field of ecological conservation and environment protection, and fully leverage the role of civil society and technology-based organizations in making scientific and technological innovation and in transforming into commercial products. Accordingly, there will be a regulating system that would conduct routine surveys on environmental protection enterprises, collect statistics, develop a credit record for companies providing environmental service, and release reports to review the development of environmental service industry.

Section 3 Drive growth through green technology innovation

Green development increasingly driven by innovation. With green development increasingly underpinning China's national innovation strategy and economic transformation, it's crucial to thoroughly incorporate green development into developing up-to-date technologies in various sectors. Intelligent green manufacturing technologies will be developed and moved up to the high end of the value chain. There will be efforts to develop green, efficient and safe modern agricultural technologies, as well as be more research and development (R&D) in such areas as water-saving, circular and organic agriculture, modern forestry and bio-fertilizer to promote high-quality, efficient and sustained development of agriculture. China will develop modern energy technologies that are safe, clean and efficient, to transform the way energy is produced and consumed. It's planned to develop technologies on resources conservation and recycling, and put in place a technology system for utilizing municipal solid waste, recycling renewable resources and comprehensively utilizing industrial solid waste. Targeting air, water and soil contamination, steps will be taken to develop a complete set of technologies to facilitate source control, end-of-pipe control and ecological restoration.

Developing a Science and Technology (S&T) innovation system for environment protection. Aiming at the forefront of world environmental S&T and fully taking the strategic requirements for environmental protection into consideration, China will move fast in establishing a national S&T innovation system on environment protection, featuring high efficiency, and a clear cut line in the responsibilities between different parts within. Such system would be able to provide strong support to environmental protection and play a crucial role in encouraging independent innovation and integrated innovation. Efforts will focus on formulating a theoretical system of S&T innovation led by scientific research, as well as a research and development (R&D) system underpinned by demonstration projects. There will be an environmental benchmark and standard system aiming to improve human health, an enabling system to develop highly competitive environmental protection industries, and a service-based S&T management system. Environmental S&T research talent development projects will be launched to cultivate more leading technical professionals and young talents in environmental field, and there will be a number of innovative talent training bases and some high-level innovation teams in place. China will support relevant universities in developing basic science and applied science on environmental protection, and establish an award system for environmental protection professionals.

Building up S&T innovation platforms for environmental protection. China will integrate all S&T resources to reform the S&T system on environmental protection. It calls for developing various S&T innovation platforms including key laboratories, engineering and technology centers, scientific observation and research stations as well as environmental think tanks. These will help develop and promote more cutting edge technologies, and enhance science-based management. Enterprises are encouraged to collaborate with the research institutions and play a primary role in innovation, so that R&D on environmental technologies will be enhanced and its achievement are able to be transferred, applied and used in a wider range of areas. There will also be supporting platforms including one information platform to inform potentials needs for environmental protection equipment and services, and one on transferring and trading of technological innovation.

There will be a number of development zones built in S&T industrial parks with suitable conditions, which include pilot zones for technology innovation, and zones for high-tech industry, high-tech on pollution treatment, international cooperation, and high-level personnel training and education. In doing so, China will establish a number of national-level high-tech industrial development zones for environmental protection.

Implementing key S&T projects on environmental protection. China will continue to implement key national projects on controlling and treating water pollution through science and technologies. A group of key R&D projects will be implemented covering technologies on tracking air pollution causes and pollution control, restoring and protecting typical fragile ecosystems, on clean and efficient use of coal and energy conservation, comprehensively preventing, controlling and remedying agricultural non-point pollution and heavy metal contamination in farmland, as well as on safeguarding marine environmental safety. There will be pilot projects in the Beijing-Tianjin-Hebei region, the Yangtze River Economic Belt, as well as provinces, autonomous regions and municipalities along the Belt and Road, where new technologies will be applied to controlling pollution and restoring ecosystems in these regions, thus helping to provide technical and systematical solutions to various environmental issues and improve environmental governance. Steps will be taken to build a community of collaborative innovation that covers Beijing-Tianjin-Hebei region and other regions where S&T innovation projects will be implemented as part of efforts to improve regional environmental quality. Innovation shall also be incorporated into developing technical methods and management models of protecting and restoring ecological-security shields in the Tibetan Plateau. There will be steps to study and develop key technologies in areas such as ecological monitoring and early warning, ecological restoration, biodiversity conservation, ecological redlining assessment and management, and ecological corridors construction. A number of demonstration areas for ecological protection and restoration technologies will be established. Support will be given to all kinds of R&D on environment monitoring and early-warning network, key technologies and equipment for monitoring ecosystems of soil, air and greenhouse gases, as well as to developing technologies on monitoring, early warning and emergency response for environmental accidents, remote sensing technologies, data analysis and service products, and high-end environmental monitoring instruments. There will also be studies to examine pollution and environmental impacts of hazardous waste discharged from key industries, to track and quickly identify hazardous waste, to guard against and control risks throughout the whole process, and to develop information management technology. It's hoped to put in place a system for assessment of environmental and health risks of chemicals, relevant procedures and for technical specifications. Scientific research will be enhanced through better decision-making on environmental management, and the research will focus on studying and applying technical methods on coordinated control of multiple pollutants, simulation of ecosystems, pollution source analysis, planning for ecological environment protection, ecological environment damage assessment, grid management and green Gross Domestic Products (GDP) accounting.

Improving environmental standards and policy systems on technologies. China will study and develop environmental baselines, revise soil environment quality standards, improve VOCs emission standards, and strictly implement pollutants discharge standards. It will work fast in developing, revising and implementing the pollutants discharge standards for motor vehicles and non-road mobile sources, and standards for petroleum quality. China will release and implement *Limits and Measurement Methods for Exhaust Pollutants from Marine Engines (phase I and II)*, *Limits and Measurement Methods for Emissions from Light-duty and Heavy-duty Vehicles (phase VI)*, *Limits and Measurement Methods for Emissions from Mopeds (phase IV)* as well as *Discharge Standards of Pollutants for Livestock and Poultry Breeding*. China will revise *Discharge Standards of Pollutants for In-service Vehicles* and strive to implement *Discharge Standards of Pollutants for Non-road Mobile Machineries (phase IV)*. Policies concerning environmental protection technologies will be improved, and technology standards to regulate ecological conservation redlining will be established. There will be efforts to improve the evaluation index system for clean production in key industries such as iron and steel, cement and chemicals. China will also expedite the development of technologies policies for key industries such as electric power, metallurgy, nonferrous metals and in major fields including urban and rural solid waste disposal, pollution control of motor vehicles and non-road mobile machinery as well as prevention and control of agricultural non-point pollution. Standards and technology system for recycling, utilization, disposal and environmentally sound processing of hazardous waste will be established.

Section 4 Promote green and coordinated regional development

Promoting green and coordinated development in four major regions. The Western China must take ecological conservation as a top priority, and double its efforts to protect the environment. Specific steps will include making ecological-security shields more able to conserve the ecosystems, developing special zones for supplying ecological products, appropriately utilizing strategic resources such as oil, coal and natural gas, and developing featured resources such as ecological tourism and agricultural and livestock products. On northeastern China, efforts will focus on protecting forest ecosystems of the Greater Khingan and the Lesser Khingan regions and Changbai Mountains, building shelterbelts, and protecting farmland soil and wetlands in the Northeast Plain, in an effort to revitalize the old industrial bases. On the central part of China, development must take full consideration of the region's resource and environmental carrying capacity, and there will be steps to orderly transfer industries, develop eco-economic zones of the Poyang Lake and Dongting Lake, and eco-economic belts along Hanjiang River and Huaihe River, develop a group of ecological corridors along river basins and traffic lines, as well as protect and improve water environment. The Eastern China will expand the ecological space, raise resources efficiency, accelerate industrial upgrading and take the lead in improving ecological environment quality.

Promoting a green Belt and Road Initiative. China will strengthen existing bilateral cooperation mechanisms with Russia, Kazakhstan, and multilateral cooperation mechanisms of China-ASEAN cooperation and Shanghai Cooperation Organization. China will actively engage in protecting environment in Lancang-Mekong River basins, and strengthen exchanges with environment officials, scholars, young people from countries along the routes to develop an all-dimensional and multi-channel exchange mechanism. China will initiate environmental campaigns for public interest and implement a plan of "Green Silk Road Envoys" to share China's visions, practice and experience in promoting ecological conservation and green development. China will establish and improve a sound management system for green investment and trade, and implement the *Guidelines for Environmental Protection in Foreign Investment and Cooperation.* Technological cooperation parks and demonstration sites to promote overseas operation of Chinese environmental industries will be set up. A number of green brands and high-quality industrial capacity on railway, power, automobile, communications, new energy and iron and steel will be developed and reach out to the world. Industrial upgrading and innovations in provinces, autonomous regions and municipalities along the routes will be promoted, and the green industrial chain will be extended. Key strategies and projects must first go through environmental impact assessment to get more prepared for potential environmental risks. Accordingly, a national plan on protecting ecological environment along the routes will be set in place.

Enhancing coordination in protecting the Beijing-Tianjin-Hebei region. Development in this region must be based on resource and environmental carrying capacity, making efforts to transform its economic growth model and optimize spatial layouts based on ecological functions to increase its environmental carrying capacity and expand its ecological space. Traditional manufacturing industry in Tianjin will be transformed towards a green model at an early date, and part of the city functions in Beijing that are thought not essential to a capital city will be gradually transferred to Hebei Province. Scientific and technological achievements will be applied and promoted in Beijing and Tianjin. Regional environmental cooperation will be strengthened by jointly addressing air, river and lake pollution. Ecological-security shields will be strengthened through jointly building Bashang plateau ecological environment protection zone and Yanshan-Taihang Mountain ecological conservation zones. Broader application of new energy sources such as photovoltaic energy will be encouraged. Innovations will be made to develop new mechanisms for coordinated management of ecological environment, including developing an integrated regional environmental monitoring network, networks for sharing ecological environment information, an early warning and emergency response system for environmental accidents, a regional mechanism for coordinated ecological environment protection, a system for water resource allocation, trans-regional supervision and law enforcement mechanism, a regional eco-compensation mechanism and an inter-regional emission trading market. It's planned that by 2020 all these mechanisms for coordinated environmental protection in Beijing-Tianjin-Hebei region are able to operate effectively and ecological environment quality in this region will be improved remarkably.

Protecting the environment in the Yangtze River Economic Belt. China will prioritize protecting and restoring the ecological environment of the Yangtze River through promoting ecological civilization and building a green ecological corridor there. There will be a holistic approach coordinating the development between water resources, water environment and water ecology, between the upper, middle and lower reaches of the river, and measures strengthening interaction and collaboration between eastern, central and western regions. Cross-sector and trans-regional regulation and coordination on emergencies response will be strengthened. It shall be clear that the Yangtze River Economic Belt is refrained from any large-scale development activities, but instead must prioritize environmental protection by implementing a number of key projects on ecological restoration. A variety of ecological elements in lakes and rivers must be managed in a coordinated manner, which calls for a sound ecological security system comprised of mainstream and tributaries of the Yangtze River as the main frames as well as hills, waters, forests, and farmlands as a whole. In this system, lakes and rivers are in sound interaction with good water quality and strengthened ecological functions, soil and water are effectively conserved, and biodiversity is well maintained. In the upstream region, protection will focus on conserving and properly utilizing water sources, conserving soil and protecting biodiversity, and attention must be paid to any potential impacts on ecosystems in developing hydropower. In the middle reaches, priority is given to fostering a sound river-lake interaction, and ensuring water safety in Danjiangkou Reservoir. In the lower reaches, there will be intensified efforts to restructure and upgrade industries, restore the degraded aquatic ecosystems, protect drinking water sources, control urban expansion which might take up more ecological space, and tackle water pollution in the river networks. Relations between rivers and lakes must be properly balanced, which calls for steps to take coordinated approach to allocate water resources in the Yangtze River mainstream and four major rivers in the upstream of the Dongting Lake and five major rivers in the upstream of Poyang Lake, to ensure the water level and water flow are maintained at certain level that could maintain its ecological functions. It's also required to formulate a holistic plan to regulate the use of resources along the Yangtze River coastal line, and control the development intensity there. There will also be cross-sectional assessment of quality of water sections to promote coordinated governance.

Chapter 4 Promote Quality-oriented Management in Implementing the Three Action Plans on Prevention and Control of Air, Water and Soil Pollution

Centering on the improvement of environmental quality, China will promote joint prevention and control of pollution and coordinated management of river basins, and develop roadmaps for implementing the action plans on prevention and control of air, water and soil pollution. As conditions in regions, watersheds and different types of pollutants are differing from one another, there will be differentiated policies for different regions and coordinated control of multiple pollutants to make the efforts more targeted and effective. Bottom lines of environmental quality will be drawn up, and based on which, phased targets will be set for improving environmental quality, and a checklist of environmental governance responsibilities will be developed and implemented, so as to address prominent environmental problems of public concern.

Section 1 Improve ambient air quality in line with regional conditions

Implementing the objective-oriented management and deadline-based planning for ambient air quality improvement. All provinces, autonomous regions and municipalities should have an overview and thorough analysis of the status quo and the development trend of air quality based on national ambient air quality standards, as well as assess the progress and make public relevant information on a regular basis. They should also strengthen process management when phasing out excess capacity in industries with heavy pollution such as steel and cement to meet relevant standards. Use of clean energy will be strongly promoted by upgrading standards for motor vehicles and fuel products, and by enhancing quality control of oil and energy products. Pollution control of mobile sources will be intensified, devoting greater efforts in the control of urban dust, scattered sources from micro and small-sized enterprises and domestic sources. China will further carry out the *Action Plan on Prevention and Control of Air Pollution*, with the measures of significantly reducing SO_2, NO_x and PM emissions, preventing and controlling VOCs pollution, and piloting ammonia emission control. It's hoped through all these efforts, China will be able to reduce SO_2 and CO concentration to a standard level in all cities at the prefecture level or above, to put a significant dent in concentration of PM2.5 and PM10, to cut NO_2 concentration on a continuous basis in cities at the prefecture level or above, and to stabilize or lower ozone concentration level. China will implement an objective-oriented management system for urban ambient air quality. Cities meeting national air quality standards should strengthen protection to maintain good momentum, and those yet to reach the standards should set a deadline for attaining the standard, release it to the public, and develop a plan that specifies the timeline, roadmap and major tasks.

Improving the response mechanism to heavy air pollution. There will be efforts to better operate and manage the centers of air quality forecast at all levels to make reliable and timely air quality forest, and to enable internet access to relevant information nationwide. China will improve joint early warning mechanism at regional level for very unhealthy or hazardous days, and strengthen the capability of ambient air quality forecast in the northeast, northwest, central regions and the Chengdu-Chongqing region. Emergency response mechanism will be improved by developing technical standards for assessing the plan for emergency response to heavy pollution, and strengthening inspection and assessment on the implementation of the mechanism. All provinces, autonomous regions, municipalities and cities at the prefecture level or above should prepare and revise emergency response plans in time. They should conduct analysis on causes and sources of pollution in order to develop scientific and targeted abatement measures. The list of emergency measures should be renewed every year. All provinces, autonomous regions and municipalities should initiate their emergency response mechanisms in a timely manner to effectively respond to heavy air pollution. Regulation and supervision will be strengthened, and local governments failing to make timely or adequate response will be called for an interview, notified, and ordered to correct their wrongdoings under supervision.

Strengthening regional cooperation in preventing and controlling air pollution. A regional cooperation mechanism will be introduced to prevent and control air pollution in the Beijing-Tianjin-Hebei region and the surrounding areas, the Yangtze River Delta region and the Pearl River Delta region. Under this mechanism, regional cooperation will be carried out on a regular basis subject to unified standards, planning, monitoring and control. There will be unified environmental standards and consistent policies for levying pollutant discharge and for energy consumption in regulating key industries and regions. There will also be unified standards in place in phasing out old vehicles and managing in-use vehicles. China will strictly control the total coal consumption in key areas and try to realize a negative growth of coal consumption in the Beijing-Tianjin-Hebei region, the Yangtze River Delta region, the Pearl River Delta region, and Shandong Province. Out of the top 10 cities suffering the most severe air pollution, those that have coal as the prime cause of air pollution must also see a negative growth of coal consumption. A market-based approach will be employed to phasing out old vehicles and vessels and to upgrade anti-pollution facilities and equipment at a quicker pace. There will be regulations on newly produced motor vehicles and non-road mobile machinery to ensure their emissions live up to environmental standards. China will launch Clean Diesel Campaign and strengthen the management of construction machinery, heavy-duty diesel vehicles and agricultural machines with high emissions. In key regions, diesel vehicles must be registered and subject to environmental examination and inspection, and there will be on-site environmental inspection of trucks, passenger cars and buses. The proportion of public vehicles using new energy will be raised. Basically all urban buses, where conditions permit, are expected to use new energy by the end of 2017. China will also implement policies on the management of vessels in emission control areas (ECAs) in waters of the Pearl River Delta region, Yangtze River Delta region and Bohai Rim (Beijing-Tianjin-Hebei region). All berthing vessels should prioritize the use of shore power. China will develop a monitoring network of remote sensing on vessels for their air pollutants discharge of vessels and a petroleum quality monitoring network, to carry out discharge monitoring and joint regulation of vessels in ECAs, in an effort to set in place a regulation system to ensure motor vehicles and vessels discharge as well as petroleum products live up to the environmental standards. Upgrade of petroleum quality for non-road mobile sources will be accelerated. Dust control and integrated management of urban roads and construction sites will also be enhanced.

Significantly reducing PM concentrations in the Beijing-Tianjin-Hebei region and its surrounding areas. With Beijing, Baoding and Langfang of Hebei province as the key areas of environmental protection and pollution control, efforts will be made to treat pollutants discharge from raw coal that is burned in winter for heating or cooking in households, restaurants and small industrial boilers, to control pollution in key industries, to regulate motor vehicles, and address weather incidents of heavy air pollution. The elevated sources will also be controlled and regulated to improve regional air quality. The ratio of external electric power and non-fossil energy supply will be raised. Major cities in the region will replace coal by natural gas and electricity in order to greatly reduce consumption of the said raw coal. The total coal consumption in Beijing, Tianjin, Hebei, Shandong and Henan will go down by 10% during the 13th Five-Year Plan period. A regional monitoring platform for vehicle emissions will be set in place at an early date, focusing on the control of heavy-duty diesel vehicles and high-emission vehicles. Regional PM2.5 concentration is expected to be cut significantly and ozone concentration will be stabilized by 2020.

Substantially reducing PM2.5 concentration in the Yangtze River Delta. The region will accelerate industrial transformation and phase out the production capacity that fails to meet energy and environmental standards according to law. During the 13th Five-Year Plan period, the total coal consumption will be cut by about 5% in Shanghai, Jiangsu, Zhejiang and Anhui, and coal-fired boilers with capacity lower than 35 t/h in cities at the prefecture level or above will be basically eliminated. Comprehensive prevention and control of VOCs in industries such as oil refining, petrochemical, industrial coating and printing will be facilitated. PM2.5 concentration in the Yangtze River Delta region is expected to see sharp reduction and ozone concentration is stabilized by 2020.

Air quality in the Pearl River Delta expected to meet standards first. The region will make an overall plan for prevention and control of PM2.5 and ozone pollution, and take coordinated actions to control both VOCs and NO$_x$ pollution simultaneously. There will be steps to expedite industrial transformation and upgrading and optimize energy structure in the region. Industrial parks and industrial zones will provide centralized heating and, if conditions permit, introduce large gas-fired heating boilers. Coal consumption is expected to fall by about 10% during the 13th Five-Year

Plan period. The region will focus on comprehensive prevention and control of VOCs emission in industries such as petrochemical, chemical, oil storage, transportation and sales, automobile manufacturing, ship-building (maintenance), container manufacturing, printing, furniture and footwear industries. The Pearl River Delta region will basically meet national air quality standards and see no weather incidents of heavy pollution by 2020.

Section 2 Improve water quality with targeted measures

Developing an objective-oriented system for water quality management based on control units. China will define terrestrial control units according to function-oriented zoning and administrative divisions, and establish a three-tier water management system including river basins, aquatic ecology control areas, and aquatic environment control units. Objective-oriented management for water quality in river basins will be established, with control units as the physical basis, water quality of sections as the management objective and pollution discharge permit system as the core. The monitoring network for the water quality of control units will be improved by establishing a feedback mechanism between pollution discharge of control units and water quality of sections, in order to clearly identify the responsibilities of certain control units for water quality deterioration and strictly control pollutant discharge. China will fully implement the "river chief system". Pilot projects will be initiated in river basins such as the Yellow River and the Huaihe River where the ecological flow (water level) is scientifically measured in different periods as an important reference to water resource allocation. China will thoroughly implement the *Action Plan on Prevention and Control of Water Pollution*, identifying the responsibility of control units for controlling pollution and meeting their objectives. The control units mainly polluted by stationary sources should identify major water pollutants and set discharge control targets for major pollutants exceeding the limit in the region or the river basin. Pollutant discharge permit system for water quality improvement will be implemented, which breaks down pollution control tasks to each pollution discharge organizations (including sewage treatment plants and large-scale livestock breeding units with discharge outlets) within control units. The control units mainly polluted by non-point sources (scattered sources) or those with severe water shortage should be managed by introducing incentives, strengthening regulation to ensure ecological flow and improve aquatic environment. All provinces are required to regularly release to the public their progress in attaining the water quality objectives of its control units as of 2017.

Box 2. Control Units in Need of Improvement of Water Quality in Each River Basin

1. The Yangtze River Basin (108 units)

Water quality improvement: from Grade IV to Grade III in 40 control units including Shuangqiao River control unit in Hefei, from Grade V to Grade III in 7 control units including the Wujiang River control unit in Chongqing, from Grade V to Grade IV in 9 control units including the Laihe River in Chuzhou, from worse than Grade V to Grade III in 2 control units including Jingshan River in Jingmen, from worse than Grade V to Grade IV in 4 control units including Tuojiang River in Neijiang, and from worse than Grade V to Grade V in 24 control units including the Shiwuli River in Hefei.

Pollutant concentration reduction: COD concentration of Dianchi Lake (outer lake) control unit in Kunming, ammonia nitrogen concentration in 3 control units including Nanfei River in Hefei, both ammonia nitrogen and total phosphorus concentrations in 4 control units including Zhupi River in Jingmen, and total phosphorus concentration in 14 control units including Minjiang River in Yibin.

2. The Haihe River Basin (75 units)

Water quality improvement: from Grade IV to Grade III in 9 control units including Yanghe River No.8 Bridge in Zhangjiakou, from Grade V to Grade IV in 3 control units including downstream Guishui River in Beijing, and from worse than Grade V to Grade V in 26 control units including Chaobai River in Tongzhou District.

Pollutant concentration reduction: COD concentration in 6 control units including XuanHui River in Cangzhou, ammonia nitrogen concentration in 26 control units including downstream Tonghui River in Beijing, ammonia nitrogen and total phosphorus concentrations in

3 control units including Communism Canal in Xinxiang, COD and ammonia nitrogen concentrations in Haihe River Floodgate in Tianjin, and total phosphorus concentration in Chaobai New River in Tianjin.

3. The Huaihe River Basin (49 units)

Water quality improvement: from Grade IV to Grade III in 17 control units including Guhe River in Fuyang, from Grade V to Grade III in the Dongyu River control unit in Heze, from Grade V to Grade IV in 9 control units including Xinsui River in Suqian, from worse than Grade V to Grade III in the Zhuzhaoxin River control unit in Heze, from worse than Grade V to Grade IV in the Yunliao River control unit in Xuzhou, and from worse than Grade V to Grade V in 16 control units including Yuefang Bridge over Wahe River in Bozhou.

Pollutant concentration reduction: Ammonia nitrogen concentration in 4 control units including Baohe River in Shangqiu.

4. The Yellow River Basin (35 units)

Water quality improvement: from Grade IV to Grade III in 14 control units including Yiluo River in Luoyang, from Grade V to Grade IV in 4 control units including Hulu River in Guyuan, from worse than Grade V to Grade IV in Lanhe River control unit in Lvliang, and from worse than Grade V to Grade V in 8 control units including Dahei River in Wulanchabu.

Pollutant concentration reduction: ammonia nitrogen concentration in 8 control units including Kundulun River in Baotou.

5. The Songhua River Basin (12 units)

Water quality improvement: from Grade IV to Grade III in 9 control units including Xiaoxingkai Lake in Jixi, and from worse than Grade V to Grade V in Ashi River control unit in Harbin.

Pollutant concentration reduction: COD concentration in Hulun Lake control unit in Hulun Buir, and ammonia nitrogen concentration in Kaoshan Nanlou control unit near Yinma River in Changchun.

6. The Liaohe River Basin (13 units)

Water quality improvement: from Grade IV to Grade III in 6 control units including Kouhe River in Tieling, from Grade V to Grade IV in 3 control units including Juliuhe Bridge over Liaohe River control unit in Shenyang, and from worse than Grade V to Grade V in 2 control units including Liangzi River in Tieling.

Pollutant concentration reduction: total phosphorus concentration of Hunhe River control unit in Fushun, and ammonia nitrogen concentration of Tiaozi River control unit in Siping.

7. The Pearl River Basin (17 units)

Water quality improvement: from Grade III to Grade II in 2 control units including Paili unit of Jiuzhou River in Zhanjiang, from Grade IV to Grade II in Niuwan control unit of Tanjiang River in Jiangmen , from Grade IV to Grade III in 4 control units including Jiangkoumen control unit of Jianjiang River in Maoming City, from Grade V to Grade IV in 2 control units including Zhangcun Village control unit of Dongguan Canal in Dongguan, from worse than Grade V to Grade IV in Shibi control unit of Xiaodong River in Maoming City, and from worse than Grade V to Grade V in 5 control unit including Hekou control unit of Shenzhen River in Shenzhen.

Pollutant concentration reduction: COD concentration in Qilu Lake control unit in Yuxi, and total phosphorus concentration of Xingyun Lake control unit in Yuxi.

8. Rivers in Zhejiang Province and Fujian Province (25 units)

Water quality improvement: from Grade IV to Grade III in 13 control units including Puyang River in Hangzhou, from Grade V to Grade III in 3 control units including Tingxi River in Xiamen, from, Grade V to Grade

IV in 5 control units including Nanxi River in Zhangzhou, and from worse than Grade V to Grade V in 4 control units including Jinqing Port in Taizhou.

9. Rivers in northwest China (3 units)

Water quality improvement: from Grade IV to Grade III in Bosten Lake in Bayinguoleng Mongol Autonomous Prefecture, from worse than Grade V to Grade III in Bedahe River in Jiuquan, and from worse than Grade V to Grade V in Kezi River in Kashi.

10. Rivers in southwest China (6 units)

Water quality improvement: from Grade IV to Grade III in 4 control units including Hehui River in the Dali Bai Autonomous Prefecture.

Pollutant concentration reduction: COD concentration of Yilong Lake in Honghe Hani and Yi Autonomous Prefecture, and ammonia nitrogen concentration in Xi'er River in the Dali Bai Autonomous Prefecture.

Comprehensively controlling the pollution in river basins. China will implement the *Action Plan on Water Pollution Prevention and Control in Key River Basins*. Governments and relevant departments at all levels upstream and downstream should strengthen coordination and conduct regular meeting for joint monitoring, law enforcement, emergency response and information sharing. Systematic protection should be enhanced in the Yangtze River Basin, with more efforts going to aquatic biodiversity conservation and pollution control of water channels and ports. Governments and relevant authorities in the Yangtze River Basin will carry out comprehensive control of total phosphorus concentration in the Minjiang River, Tuojiang River, Wujiang River, Qingshui River and Yichang section in the mainstream of the Yangtze River to effectively control total phosphorus pollution in Guizhou, Sichuan, Hubei and Yunnan. Comprehensive pollution control should continue in the Taihu Lake, enhancing ecosystem functions to prevent blue-green algae bloom and ensure drinking water safety. Total nitrogen and total phosphorus control should be strengthened in the Chaohu Lake, improving the quality of in-flowing water to restore lakeside ecological functions. Total nitrogen and total phosphorus control in the Dianchi Lake should be strengthened, with a focus on the prevention and control of urban sewage and inflow of agricultural non-point source pollution. Area-specific ecological restoration should be carried out by stage in order to gradually recover the aquatic ecosystem in the Dianchi Lake. For the Haihe River Basin, water conservation and recycling should be highlighted. Trans-boundary water management should be enhanced, focusing on treatment of black and putrid water in urban and rural areas to meet ecological water demand in Baiyangdian Lake, Hengshui Lake and Yongding River. Ammonia nitrogen pollution should be effectively contained in the Huaihe River Basin by significantly reducing emission intensity of industries such as paper making, fertilizer and brewage. Water quality of tributaries of the Huaihe River, such as the Honghe River, Guohe River, Yinghe River, Huiji River and Baohe River, should continue to improve. Environmental pollution emergencies should be strictly prevented. For the Yellow River Basin, control of pollution discharge from coal chemical and petrochemical enterprises should be strictly monitored. Water quality of tributaries, including Fenhe River, Sushui River, Zongpaigan River, Dahei River, Ulasu Lake, and Huangshui River, should continue to improve, minimizing aquatic environmental risks of its middle and upper reaches. Water quality of tributaries of the Songhua River, such as Ashi River and Yitong River, should continue to improve, addressing the heavy water pollution of industries such as petrochemical, brewing, pharmaceutical and papermaking. Aquatic ecological conservation should be strengthened by increasing wild fish population and accelerating restoration of wetland ecosystems. Pollution discharge intensity of industries such as petrochemical, papermaking, chemicals, and agro-food processing will be greatly reduced in the Liaohe River Basin. Through sustained improvement in water quality of its tributaries such as the Hunhe River, Taizi River, Tiaozi River and Zhaosutai River, the aquatic ecosystems will be significantly recovered and wetland ecosystems fully restored. For the Pearl River Basin, an integrated system for the prevention and control of water pollution in Guangdong, Guangxi and Yunnan will be established, to ensure good water quality of water supply of Dongjiang River and Xijiang River, and to improve aquatic ecology of the Pearl River Delta region.

Prioritizing protection of waters with good quality. China will supervise and regulate water use from source to tap to ensure sustained improvement in drinking water safety. Local governments at all levels and water suppliers should conduct regular monitoring, test and assessment on drinking water sources, water plant outflow and tap water within their administrative areas to ensure drinking water safety. All cities at the county level or above should make public the drinking water safety information on a quarterly basis and cities at the prefecture level or above should make public such information as of 2018. China will develop standards on drinking water sources and remove according to law any illegal buildings and outlets within drinking water source protected areas. China will also enhance the protection of rural drinking water sources and carry out projects on improving the safety of drinking water in rural areas. All provinces, autonomous regions and municipalities will basically finish the identification of centralized drinking water source protected areas at the county level or above by the end of 2017, and start conducting regular monitoring and assessment. Over 93% of centralized drinking water sources of cities at the prefecture level or above are expected to meet Grade III water quality standard by 2020. China will carry out assessment of ecological security of all river sources as well as rivers, lakes and reservoirs with water quality at or better than Grade III, and develop ecological conservation plans. The Dongjiang River Basin, Luanhe River Basin, Qiandao Lake Basin and Nansi Lake Basin should finish the assessment and plan development by the end of 2017. The seven key river basins will formulate and implement their plans on aquatic biodiversity conservation.

Comprehensively controlling and preventing groundwater pollution. China will conduct regular investigation and assessment of centralized groundwater drinking water sources and the surrounding areas of pollution sources. China will also strengthen regulation on groundwater environment of key industries and take effective measures to reduce risks of groundwater contamination. List of plots with groundwater pollution will be made public to contain contamination risks. Pilot pollution remediation projects will also be conducted in the polluted sites. The tendency of worsening groundwater pollution nationwide is expected to be under preliminary control, with only about 15% groundwater with very poor quality by 2020.

Treating urban black and putrid waters. China will establish a list of black and putrid waters in built-up areas of cities at the prefecture level or above. A remediation plan will be developed where milestone targets and tasks will be identified. Annual work progress and water quality improvements will be made public. China will set up a national monitoring platform for the remediation of urban black and putrid waters where the above-mentioned lists are published for public comments. All cities should disclose in key local media the information about list of black and putrid waters, deadline for meeting set standards, party or person in charge, and progress and result of control. They should establish a long-term mechanism for routine maintenance and monitoring of such waters. Black and putrid waters are expected to be basically eliminated in built-up areas of all municipalities, provincial capitals and municipalities with independent planning status by the end of 2017. All cities at the prefecture level or above are expected to eliminate all large-scale floating debris, garbage at river bank and illegal sewage outlets by the end of 2017. The proportion of black and putrid waters in built-up areas of cities at the prefecture level or above is expected to be kept within 10%, with sharp reduction of black and putrid waters in other cities by 2020.

Improving ecological environment quality of estuaries and offshore waters. China will carry out the *Plan for the Prevention and Control of Offshore Marine Pollution* and make more efforts in pollution control of offshore waters such as the Bohai Sea and the East China Sea. Direct discharge of marine pollutants and coastal industrial parks will see stricter monitoring to prevent marine pollution caused by land-based oil spills in the coastal regions. China will control ballast water and pollutants discharged by international sailing ships. The distribution of sewage outlets to the sea will be examined and all illegal and inappropriate outlets will be removed by the end of 2017. Most rivers that flow into the sea in coastal provinces, autonomous regions and municipalities are expected to reach Grade V or better water quality by 2020. China will carry out the *Comprehensive Blue Bay Improvement Plan*, with a focus on the prevention and control of estuarine pollution in the coastal zones such as the Yellow River estuary, Yangtze River estuary, Minjiang River estuary, Pearl River estuary, Liaodong Bay, Bohai Bay, Jiaozhou Bay, Hangzhou Bay and North Bay. China will adopt strict measures during fishing ban and fishing moratorium. Coastal aquaculture density will be strictly controlled to facilitate eco-friendly aquaculture, with measures of vigorously restocking aquatic organisms and promoting the development of artificial reefs and marine ranching. China will enhance conservation and restoration of coastal ecosystems, carrying out wetland restoration projects such as planting mangroves in the south and willows in the north, and strictly controlling reclamation activities in ecologically sensitive areas. National natural coastline (excluding island coastline) is expected to keep at a level no lower than 35%, with improvement and restoration of 1,000 km coastline by 2020. China will develop a group of marine nature reserves, special marine protected areas and aquatic germplasm conservation areas. China will also conduct projects of ecological protection on islands and reefs to strengthen rare marine species conservation.

Section 3 Tackle soil contamination by category

Advancing the basic research and monitoring network. China will implement the *Action Plan on Prevention and Control of Soil contamination* in an all-round way. It will carry out detailed investigation of soil contamination, with a focus on agricultural land and land use of key industries and enterprises. Area and distribution of soil contamination in agricultural land and the impact of soil contamination on the quality of agricultural products will be identified by the end of 2018. Distribution of contaminated sites in land used by enterprises in key industries, as well as their environmental risks, will be examined by the end of 2020. Investigation will be conducted on the risks in areas with many soil environment problems such as electronic waste dismantling, waste plastic recycling, informal landfill and outdated tailings ponds, and a catalogue for risk management and control will be established. China will make an overall plan to integrate and optimize the distribution of soil quality monitoring sites. Giving full play to the role of industrial monitoring network, China will support an increase in the number of sites and monitoring indicators, and frequency of monitoring based on local conditions. All national monitoring sites are expected to be set, forming a national network on monitoring soil environment quality by the end of 2017, when soil environmental monitoring capacity is basically developed. China will realize full coverage of monitoring sites in all counties, districts, and county-level cities by 2020.

Managing the environment of agricultural land by category. Based on degree of contamination, agricultural land is classified into three categories: 1) Prioritized for Protection, referring to land with no or slight contamination; 2)

Safe to Use, referring to land with slight or intermediate contamination; and 3) Strictly Controlled, referring to land with heavy contamination. Provincial governments should give early warning to counties, districts and county-level cities with reduction in arable land classified as Prioritized for Protection or degradation of agricultural land quality. They should also take restrictive measures according to law such as rejecting environmental impact assessment (EIA) of any new construction project. All arable land classified as Prioritized for Protection will be designated as permanent basic farmland, under strict protection from any reduction or degradation. Considering local soil contamination and extent of produce exceeding limits of hazardous substances, the counties, districts and county-level cities with arable land classified as Safe to Use should develop and implement a plan for safe use of contaminated farmland based on main crop varieties and local cultivation habits, taking measures such as agronomic control and crop substitution to minimize risks of produce exceeding the limits. Use of arable land classified as Strictly Controlled will be under strict control. The areas prohibited from producing certain agricultural produce will be identified according to law, where planting of edible agricultural products is strictly prohibited. Pilot project on the remediation of farmland with heavy metal contamination and adjustment of crop mix in Changsha-Zhuzhou-Xiangtan region of Hunan Province will continue. Area of heavily contaminated farmland where crop mix adjustment or Grain for Green Project is carried out is expected to reach 20 million mu by 2020.

Better managing and controlling environmental risks in construction land. China will establish a compulsory soil quality investigation and assessment system for construction land. Management system and policies will be established which integrate analysis of soil quality status, remediation of contaminated sites and redevelopment of land. As of 2017, land tenure holders will be responsible for investigation and assessment of the soil quality of the land used by industries such as non-ferrous metal smelting, petroleum processing, chemicals, coking, electroplating and leather, and the land the tenure of which is to be reclaimed or the use of which is to be changed into public facilities such as residential and commercial buildings, schools, health care and elderly care institutions. For any such land the tenure of which has already been reclaimed, local municipal or county governments shall be responsible for conducting such investigation and assessment. China will incorporate soil environment management requirements into urban planning and land management and land development, which emphasizes that land use and development must meet soil quality requirements. For contaminated sites not to be developed or yet to meet conditions for remediation at the moment, local county governments shall identify the control area, set up clear signs on the above identified areas and issue public notice. They should also be responsible for environmental monitoring on soil, surface water, groundwater and air.

Controlling and remediating soil contamination. Targeted on typical contaminated agricultural land and contaminated sites, China will carry out 200 pilot projects on soil contamination control and application of remediation technologies in batches in order to accelerate the development of a sound technical system. As of 2017, all provinces, autonomous regions and municipalities will gradually establish lists of contaminated sites and negative lists for the development and use of the contaminated sites in order to properly identify land use. Cities with densely distributed contaminated sites, including cities in the Beijing-Tianjin-Hebei region, Yangtze River Delta region, Pearl River Delta region and the industrial bases in northeast China and those with exhausted mineral resources, should carry out pollution control and remediation of contaminated sites and the redevelopment and utilization of them in a standard and organized manner. The provinces, autonomous regions and municipalities with densely distributed contaminated farmlands, such as those in the middle and lower reaches of the Yangtze River, Chengdu Plain, and Pearl River Basin, should develop and implement their plans on farmland co control and remediation by the end of 2018. China plans to release the Measures on Lifetime Accountability for Soil contamination Control and Remediation of Contaminated Land by the end of 2017. China will establish whole-process supervision system for soil contamination control and remediation, which requires strict examination on remediation plans, strengthened supervision and inspection on remediation process, and third-party assessment of remediation performances.

Controlling and preventing soil contamination in key areas. Focusing on the contaminated sites resulted from the national policy of "reducing the percentage of secondary industry and raising that of tertiary industry" in the Beijing-Tianjin-Hebei region, development and use of construction land will be strictly controlled to prevent soil environmental risks. Monitoring and supervision of soil environment in sewage irrigation areas and areas with intensive facility agriculture will be enhanced. The black soil region in northeast China will be protected in a more powerful manner, with comprehensive control measures such as reuse of straw in farmland, more application of organic fertilizer, and crop rotation and fallow. With a focus on contaminated sites left by highly-polluting enterprises and industries such as chemicals, electroplating, printing and dyeing, environmental supervision on the development and use of contaminated land will be strengthened. Facing the pollution of heavy metals such as cadmium and arsenic, the Xiangjiang River basin will see measures on contaminated farmland, such as agronomic control, adjustment of crop mix and grain for green, Risks of agricultural produces exceeding limits for hazardous substances will also be strictly controlled. For the southwest region, the focus of risk control will be the pollution generated during the process of exploiting mineral resources such as non-ferrous metals and phosphorus. Soil contamination caused by the past mining and mineral processing of phosphorus, mercury and lead will be addressed. In addition, China will launch the development of pilot regions on the comprehensive prevention and control of soil contamination in six areas, namely Taizhou (Zhejiang Province), Huangshi (Hubei Province), Changde (Hunan Province), Shaoguan (Guangdong Province), Hechi (Guangxi Autonomous Region), and Tongren (Guizhou Province).

Chapter 5 Promote Up-to-Standard Discharge and Emissions Reduction While Targeting Special Sectors and Areas

With up-to-standard discharge as the bottom line and implementation of key projects as the starting point, China will improve the total emissions control system, and facilitate multi-pollutant control and reduction with coordinated actions between urban and rural areas. Increment of emission will be strictly controlled, and stock of pollutants will be sharply reduced to mitigate pressure on the ecological environment.

Section 1 Implement plans on up-to-standard discharge from all industrial pollution sources

Conducting self-monitoring and information disclosure for all industrial pollution sources. Industrial enterprises should establish an environmental management accounting system, perform self-monitoring and truthfully report their pollution discharge. Major pollutant discharging organizations should disclose relevant information according to law. China will strengthen standards for discharge outlet management. All industrial enterprises must set up their outlets in line with the standards, and prepare annual reports on pollutant discharge by the end of 2018. All pollutant-discharging enterprises will conduct comprehensive on-line monitoring. Local governments at all levels should improve the warning mechanism for excessive and abnormal discharge by major pollutant-discharging organizations, and gradually develop a unified collection and release process for monitoring discharge data from industrial sources, which facilitates strengthened public supervision under which law compliance and commitment fulfillment of enterprises can be examined. A national environmental regulatory information platform for industrial enterprises will be established by the end of 2019.

Investigating and releasing the list of blow-standard industrial pollution

sources. All provinces, autonomous regions and municipalities will enhance supervision and inspection of industrial pollution sources. A "dual random" inspection system will be promoted, where inspected subjects and inspectors are selected in a random manner. A color system for environmental credit evaluation will be developed to explore a quantitative management approach where excessive discharge by enterprises will be scored. Any enterprise failing to meet discharge standards or exceeding discharge limits for major pollutants will be warned with a "yellow card" and restricted or suspended for production. Those that fail to meet requirements after improvement and that seriously violate relevant laws and rules will be given a "red card" and shut down for a given period of time. Local governments at all levels should develop a plan for up-to-standard discharge of all industrial pollution sources within their respective administration and set annual objectives as of 2017. The list of enterprises with "yellow card" or "red card" should be published quarterly. Ministry of Environmental Protection (MEP) will devote more efforts into random check and examination, and will notify and call for public supervision on governments of those regions where enterprises exceeding discharge standards are common and concentrated.

Setting a deadline for enterprises in key industries to meet discharge standards. China will establish a mechanism for open selection and promotion of practical technologies on pollution control by industry, and publish pollution control technologies in key industries. Deadline-based improvement plan will be developed and implemented in key industries by river basin or by region. Enterprises are required to upgrade environmental protection facilities. Strict inspection and examination will be conducted to ensure continued compliance with environmental standards. In addition, China will facilitate reform of enterprises to meet discharge requirements in key industries such as steel, cement, petrochemicals, non-ferrous metals, glass, coal-fired boilers, papermaking, printing and dyeing, chemicals, coking, fertilizer, agro-food processing, bulk pharmaceutical chemicals (BPC) manufacturing, leather, pesticides, and electroplating.

Improving centralized sewage treatment facilities in industrial parks. Wastewater in industrial parks will be collected and disposed separately by category and quality, "separating sewage from clean water and rainwater". Enterprises should meet discharge requirements before entering an industrial park, and should connect to the centralized wastewater treatment facilities after entry. The final outlet of the centralized sewage treatment facilities in the industrial park should be equipped with automatic monitoring and video monitoring system, and connected to the networks of environmental protection authorities. China will conduct demonstration projects on developing standardized operations of centralized sewage treatment in industrial parks.

Section 2 Further reduce major pollutants discharge

Improving the total emissions control system. China will develop a differentiated management system at all levels which identifies requirements for total emissions control, with improving environmental quality as the core and major abatement projects as the starting point. The accounting system for total emissions reduction will be optimized, with province as the main unit of accounting. China will promote a voluntary abatement management system where sustained and effective measures for environmental quality improvement are encouraged to be included in reduction accounting. China will enhance the control of major environmental protection projects and give early warning to any region lagging behind schedule. Local pollution reduction projects and indicators should be made public. Examination on the total emissions reduction will be subject to the examination on environmental quality. Areas with below-standard environmental quality or obvious inconsistency between pollution reduction data and the trend of environmental quality will be under close scrutiny. Based on the findings of environmental supervision, routine inspection and implementation of pollutant discharge permit system, China will carry out "dual random" inspection on the pollution reduction management of all provinces, autonomous regions and municipalities. In addition, China will strongly promote total emissions control in all regions and industries. Regional and industrial amount control of typical pollutants should be strongly promoted, and incorporated into local economic and social development plans.

Promoting pollution control and emission reduction projects. All provinces, autonomous regions and municipalities should develop special pollution control plans for 10 key water-related industries, such as papermaking, printing and dyeing, in order to substantially cut pollution discharge. Water-intensive industries, including electric power, steel, textile, papermaking, petroleum, petrochemicals, chemicals and food fermentation, should meet stricter standards and quota. Focusing on transformation of coal-fired plants for ultra-low emission, China will conduct comprehensive management and synergic control of multiple pollutants, such as SO_2, NO_x, smoke and dust and heavy metals, in key industries including electric power, steel, building materials, petrochemicals and non-ferrous metals. All provinces, autonomous regions and municipalities will develop and publish special pollution control plans by the end of 2017. Projects failing to meet pollution control requirements should be made public. Moreover, China will develop industry-specific policies on pollution control technology, and develop demonstration enterprises and projects.

Box 3. Pollution Control and Emission Reduction in Key Industries

1. Papermaking industry

Vigorous measures will be taken to replace the original paper bleaching technologies with non-chlorine or other less polluting options. Technologies on the bio-treatment process of water after primary treatment will be updated and those on the tertiary treatment process will be developed. Central control system will be improved.

2. Printing and dyeing industry

Dyeing process will be transformed to reduce drainage, and wastewater generated from the process will be comprehensively recycled. Sewage will be separated from clean water, and water with different pollutant concentration will be separated for treatment and recycling. Technologies on the bio-treatment process for water after primary treatment will be improved, and those on the tertiary treatment process such as strong oxidation and membrane treatment will be developed.

3. Monosodium glutamate (MSG) industry

Wastewater recycling will be improved by adopting measures such as flocculation, air flotation and evaporation to the wastewater and ion-exchange MSG wastewater. Wastewater will be treated with anaerobic-aerobic secondary bio-treatment process before discharge, and tertiary treatment before discharge in sensitive areas.

4. Citric acid industry

Recycling technology will be applied for low-concentration wastewater treatment, and spray granulation for high-concentration wastewater.

5. Nitrogen fertilizer industry

Technologies on hydrolysis analysis of process condensation will be transformed, and wastewater containing cyanide and ammonia will be comprehensively treated.

6. Alcohol and beer industry

Physiochemical-biochemical process will be applied for low concentration wastewater which will be collected and treated by industrial parks after pretreatment. The beer industry will apply on-site cleaning technologies.

7. Sugar industry

Recycling and comprehensive utilization of molasses and distillery effluents will be promoted by applying vacuum suction filtration machine without filter gauze, pressure washing, dry handling of sugar beets, and recycling of wastewater generated from pressed sugar beet pulp. Bio-treatment and recycling of wastewater is strongly encouraged, and discharge limit will be set in sensitive areas.

8. Starch industry

Anaerobic + aerobic bio-treatment technologies will be adopted, and an on-line monitoring and control system will be established for wastewater treatment facilities.

9. Slaughter industry

Pre-treatment of wastewater to be discharged will be enhanced, and discharge limit will be set in sensitive areas. Membrane bioreactors will be used for tertiary treatment where conditions permit.

10. Phosphorus chemical industry

Technologies on purifying wet process phosphoric acid (WPA) will be transformed to prohibit new production capacity of superphosphate and calcium magnesium phosphate fertilizer. Exhaust gas generated by furnaces of phosphorus production will be purified and recycled for synthetic organic chemical products. Incorporation of phosphorus slag and phosphogypsum in various building materials and additives is encouraged.

11. Coal-fired power industry

Ultra-low emission and energy-saving of coal-fired power plants will be promoted. Dust control measures will be implemented for open coal yards, and the open coal yards where conditions permit will be closed off for upgrading.

12. Iron and steel industry

Technical transformation of coke dry quenching (CDQ) will be conducted, and different types of wastewater will be pre-treated separately. Any sintering machine and pellet production equipment not included in phase-out plan will apply flue gas desulfurization, and flue gas bypasses are prohibited. Head and tail of sintering machines, coke ovens, blast furnaces, and converter gas dust removal facilities will be upgraded. Open material storage yards will be closed off for upgrading. A closed belt corridor for feedstock transport facilities will be installed. Transfer stations and material unloading points will be equipped with ventilation equipment.

13. Building materials industry

Stockyard and transport equipment in all processes such as raw material crushing, production, transportation, and handling to effectively control fugitive emissions. Flue gas denitrification will be carried out for all cement kilns. Cement kilns and in-line kiln/raw mills will be transformed to improve its efficiency in dust removal. "Coal to gas" and "coal to electricity" projects will be conducted in the flat glass industry. Mixture and burning of inferior materials such as high-sulfur petroleum coke is strongly prohibited. All float glass production lines not using clean energy will conduct flue gas desulfurization. All float glass production lines will carry out efficient removal of dust and nitrogen in its flue gas. Production of building ceramics and sanitary ware will employ clean fuels, with desulphurization and dust removal facilities installed in spray drying towers and ceramic kilns. Denitrification measures will be taken to any spray drying towers failing to stably meet NOx emission standards.

14. Petrochemical industry

Flue gas treatment will be carried out for catalytic cracking units, among which those fail to stably meet standards of sulfur exhaust will recycle the exhaust to improve sulfur recovery rate or install desulphurization facilities.

15. Non-ferrous metal industry

Collection of surplus flue gas will be strengthened. Double-absorption process will be employed for flue gas with SO_2 concentration higher than 3.5%. Desulfurization must be conducted for low concentration flue gas and exhaust produced during sulfuric acid production that exceeds the limits. Setup of flue gas stack by smelting enterprises will be performed by standards, and all bypasses of desulphurization facilities will be demolished.

Adopting measures to control VOCs emission in key industries and key areas. China will strictly control VOCs emission in key industries such as petrochemicals, organic chemicals, surface coating, packaging and printing. Provinces with serious PM2.5 and ozone pollution will carry out total emissions control in relevant industries. They should also develop specific targets and implementation plans to control the total amount of VOCs. China will strengthen synergic reduction of VOCs and NOx, with focused actions on aromatic hydrocarbons, olefin hydrocarbons, acetylenes, aldehydes and ketones, by developing a list of emissions from fixed sources, mobile sources and non-point sources. China will carry out a special campaign of "Leak Detection and Repair" in the petrochemical industry, taking measures against fugitive emissions. All localities should set the deadline for completing vapor recovery of gas stations, oil depots and tankers with a recovery rate of no less than 90%. Vapor recovery of crude oil or refined oil terminals should also be accelerated. For the coating industry, alternative paints with low VOCs will be used. Coating processes and relevant facilities will be improved. VOCs collection and treatment facilities will be built. The printing industry will replace all its raw material with those with low VOCs emission and improve its production technologies. Cities in the Beijing-Tianjin-Hebei region and the surrounding areas, Yangtze River Delta region, Pearl River Delta region, Chengdu, Chongqing, Wuhan and their surrounding areas, central part of Liaoning, Guanzhong region in Shaanxi, as well as Changsha-Zhuzhou-Xiangtan will strengthen VOCs emission control in an all-round way.

Implementing total emissions control in river basins or regions exceeding their respective discharge limits on total phosphorus and total nitrogen. Control units failing to meet discharge limit on total phosphorus and relevant regions upstream should control their total phosphorus amounts, defining clear indicators as binding targets and developing a water quality improvement plan. China will focus its efforts on selecting 100 phosphate mines and on developing and transforming production processes and sewage treatment facilities of phosphorus chemical enterprises. China will greatly promote recycling and reuse of ammonium phosphate wastewater and comprehensive utilization of phosphogypsum to ensure a phosphorus recovery rate of over 96% in phosphoric acid manufacturers. Total nitrogen amount control will be applied in coastal cities at the prefecture level or above and in rivers flowing into eutrophic lakes and reservoirs. Sources of total nitrogen pollution will be investigated, in order to identify key regions, fields and industries for pollution control and development of total nitrogen control plans. Total nitrogen should also be incorporated into regional targets of total emissions control. Industries such as nitrogen fertilizer and MSG should improve efficiency in supplementary material use and make more efforts in recycling. Urea consumption should be reduced or replaced with alternatives in the printing and dyeing industry. Targeted management of wastewater treatment facilities should be accelerated in industries such as papermaking where the application of nutrient salts is strictly controlled. Biological process for treating phosphorus and nitrogen will be strengthened in urban sewage treatment plants to realize synergic control of total phosphorus, total nitrogen, COD and ammonia nitrogen in the livestock and poultry industry.

Box 4. Areas under Total Emissions Control based on Regions or River Basins

1. VOCs

VOCs control will be applied in 16 provinces and municipalities with serious PM2.5 and ozone pollution, namely Beijing, Tianjin, Hebei, Liaoning, Shanghai, Jiangsu, Zhejiang, Anhui, Shandong, Henan, Hubei, Hunan, Guangdong, Chongqing, Sichuan and Shaanxi.

2. Total phosphorus

Total phosphorus control will be applied in the control units exceeding discharge limits of total phosphorus and relevant areas upstream, namely Baodi District in Tianjin, Jixi in Heilongjiang, Qiannan Buyi and Miao Autonomous Prefecture and Qiandongnan Miao and Dong Autonomous Prefecture in Guizhou, Luohe, Hebi, Anyang and Xinxiang in Henan, Yichang, Shiyan, Changde, Yiyang and Yueyang in Hubei, Nanchang and Jiujiang in Jiangxi, Fushun in Liaoning, Yibin, Luzhou, Meishan, Leshan, Chengdu and Ziyang in Sichuan, Yuxi in Yunnan.

3. Total nitrogen

Total nitrogen control will be applied in 56 coastal cities or regions at the prefecture level or above, including Dandong, Dalian, Jinzhou, Yingkou, Panjin, Huludao, Qinhuangdao, Tangshan, Cangzhou, Tianjin, Binzhou, Dongying, Weifang, Yantai, Weihai, Qingdao, Rizhao, Lianyungang, Yancheng, Nantong, Shanghai, Hangzhou, Ningbo, Wenzhou, Jiaxing, Shaoxing, Zhoushan, Taizhou, Fuzhou, Pingtan Comprehensive Experimental Zone, Xiamen, Putian, Ningde, Zhangzhou, Quanzhou, Guangzhou, Shenzhen, Zhuhai, Shantou, Jiangmen, Zhanjiang, Maoming, Huizhou, Shanwei, Yangjiang, Dongguan, Zhongshan, Chaozhou, Jieyang, Beihai, Fangchenggang, Qinzhou, Haikou, Sanya, Sansha, and county-level divisions directly under Hainan Province.

Total nitrogen control will be applied in the catchments of 29 eutrophic lakes and reservoirs, including Chaohu Lake and Nanyi Lake in Anhui, Longgang Lake in Anhui and Hubei, Huairou Reservoir in Beijing, Qiaoshui Reservoir in Tianjin, Baiyangdian Lake in Hebei, Songhua Lake in Jilin, Hulun Lake and Ulsu Lake in Inner Mongolia, Nansi Lake in Shandong, Baima Lake, Gaoyou Lake, Hongze Lake, Taihu Lake and Yangcheng Lake in Jiangsu, West Lake in Zhejiang, Dianshan Lake in Shanghai and Jiangsu, Dongting Lake in Hunan, Gaozhou Reservoir and Hedi Reservoir in Guangdong, Luban Reservoir and Qionghai Lake in Sichuan, Dianchi Lake, Qilu Lake, Xingyun Lake and Yilong Lake in Yunnan, Shahu Lake and Xiangshan Lake in Ningxia, and Ebinur Lake in Xinjiang.

Section 3 Advance infrastructure construction

Improving the urban sewage treatment systems. China will strengthen the construction of urban sewage treatment and pipelines, making more efforts to transform pipelines for separating rainwater and clean water from sewage. Construction of sewage interception, collection and piping facilities will be prioritized in urban villages, old downtown area and rural-urban fringes, in order to eliminate river water intrusion and groundwater infiltration. All counties and key towns are expected to have the capacity of sewage collection and treatment by the end of 2020, with sewage treatment rate at 95% for cities, 85% for counties, and 100% collection and treatment of sewage in almost all of the built-up areas of cities at the prefecture level or above. Sewage recycling and sludge disposal will be improved. Sludge will be degraded and recycled in an environmental-friendly manner, with environment-friendly treatment rate of 90% in cities at the prefecture level or above and 95% in the Beijing-Tianjin-Hebei region. Rainwater pollution needs to be controlled at the primary stage. Rainwater flowing into natural water bodies must be purified by the coastlines. Construction and improvement of interception pipes will be accelerated along the coast, to control leakage and combined sewer overflow (CSO). Based on local conditions, China will improve urban black and putrid waters through multiple scientific measures such as river-specific approach, source control, pollution interception, and treatment of internal sources. Urban sewage treatment plants should be upgraded according to specific circumstances. A wetland ecological treatment system should be introduced where conditions permit, to recycle wastewater for resources and energy. In environmentally sensitive areas (major lakes, major reservoirs and catchments of coastal waters), all urban sewage treatment facilities should meet Grade I-A discharge standard by the end of 2017. Newly-developed urban sewage treatment facilities in built-up areas of cities failing to meet Grade IV standards for surface water must comply with Grade I-A discharge standards. Utilization of reclaimed water is expected to reach 20% in cities facing water shortage and 30% in the Beijing-Tianjin-Hebei region by 2020. Sanitation facilities and sewage treatment facilities in ports and plants for ship building and repair will be incorporated into urban planning to improve the capacity in the treatment of oily wastewater, chemical tanker wash water and sewage. Ballast water management will be promoted.

Fully disposing urban waste and ensuring urban waste disposal facilities under stable up-to-standard operation. China will expedite the building of garbage treatment facilities in all counties to achieve full coverage of urban waste disposal facilities. Municipal solid waste generated in urban areas is expected to see reduction in amount, and will be recycled and reused to a larger extent, with over 95% of the waste disposed in an environment-friendly manner, and over 90% of the villages have their municipal solid waste under effective management. Large and medium-sized cities will focus on development of waste-to-energy incineration technologies, and will encourage joint development and sharing of incineration facilities across regions. Bio-treatment technologies will be actively developed and landfill techniques properly employed. Waste incineration rate is expected to reach 40% by 2020. China will improve waste handling system. Cities at the county level or above will realize enclosed waste collection and separate transport of dry and wet waste. Treatment and disposal of leachate and incineration ash, use of landfill methane and its odor treatment will be strengthened, and pollutant discharge of waste disposal facilities will be made public. A recycling and environment-friendly treatment system will be developed for urban kitchen waste, construction waste and textile waste. China will develop model cities (or districts) on garbage classification and demonstration projects on municipal solid waste management, and build kitchen waste treatment facilities in large and medium-sized cities. In addition, China will support co-processing of municipal solid waste in cement kilns.

Promoting the development of sponge cities. The ideas for urban planning and construction should be updated to protect and restore urban ecology. Overall regional management will be promoted in old downtown areas with a problem-oriented approach, avoiding large-scale demolition and construction. Specific issues such as waterlogging, collection and use of rainwater and treatment of black and putrid waters must be first addressed. Newly-developed urban areas will prioritize ecological environment protection to realize the set targets, and properly control the development intensity. The design of buildings, residential areas, roads, squares, parks, green space, rainwater storage and drainage facilities needs to take into consideration the function of "sponge", facilitating water infiltration, retention, storage, purification, utilization and discharge. China will greatly promote the development of up-to-standard urban drainage facilities and accelerate transformation and elimination of waterlog-prone sites. Urban built-up areas where 70% of the rainfall can be absorbed or used on site are expected to reach more than 20% by 2020. Urban water conservation will be improved. Water-saving devices must be employed in all public buildings and will be encouraged in urban households. All water-shortage cities at the prefecture level or above will meet national standards for water-efficient city by 2020, and cities in the Beijing-Tianjin-Hebei region, Yangtze River Delta region and Pearl River Delta region will meet such standards by 2019.

Increasing the supply and use of clean energy. China will prioritize the grid integration of hydropower and clean energy projects such as wind, solar and biomass energy included in the National 13th FYP for Energy Development, and implement government-subsidized full-price acquisition policy for renewable energy. Non-fossil energy will take up 39% of the total installed capacity, and the proportion of coal in total energy consumption will go down to 58% or less by 2020. China will expand the scope where consumption of high-pollution fuel is forbidden and urban gas utilization rate is raised. Untreated coal will be prohibited in areas covered by urban heating and gas network in cities at the prefecture level or above. Key regions and cities such as the Beijing-Tianjin-Hebei region, Yangtze River Delta region and Pearl River Delta region will carry out "coal to gas" project. Substitution of untreated coal in rural areas of north China will be promoted. In addition, China will accelerate the construction of charging facilities for new energy vehicles. Government agencies and large and medium-sized institutions and enterprises should take the lead in installing supporting facilities. Promotion of new energy automobiles will continue.

Promoting clean coal utilization. China will strengthen quality management of commodity coal, restricting the exploitation and sales of high-sulfur and high-ash coal resources. Coal washing and processing will be promoted, with expected coal washing rate of over 75% by 2020. China will vigorously promote the substitution of coal by electricity, gas and other clean energy sources. Regions yet to reach conditions for replacing coal with clean energy sources will actively employ clean coal to replace original coal resources. China will develop clean coal distribution centers, and establish fully enclosed coal distribution centers in all counties (or districts), as well as a clean coal supply network covering all towns and villages. Transformation of steam-condensing power generator sets (which only generates power, not heat) will be accelerated, and co-generation units will replace small coal-fired boilers to promote urban central heating. All smaller coal-fired boilers with capacity less than 10 t/h will be phased out, except for those that must be retained, in built-up areas of cities at the prefecture level or above by 2017.

Section 4 Tackle agricultural pollution and improve rural environment in a comprehensive manner

Comprehensively improving the rural environment. China will carry out in-depth patriotic health campaign and promote urban-rural environmental health campaign in order to build a healthy, livable and beautiful homeland. While further implementing the policy of "employing award to facilitate pollution control", China will promote a new round of continuous improvement on rural environment mainly in areas surrounding important water sources such as the areas along the routes of South-to-North Water Diversion Project, the Three Gorges Reservoir and the areas along the Yangtze River. Provincial governments where conditions permit should conduct full-coverage environmental improvement. The mode of municipal solid waste disposal in rural areas should be improved by encouraging on-site recycling and the mode of "collection in villages, transport in towns and disposal in counties". The problems of villages and dams surrounded by garbage will be addressed to prevent the shift of urban waste to rural areas. All counties should promote unified planning, construction and management of rural sewage treatment within their respective administration. Urban sewage and waste treatment facilities and services will be extended to rural areas, and toilets will be made environmentally friendly in rural areas. China will continue its program on development of clean rural areas and conduct river dredging projects. By 2020, 130,000 administrative villages will complete comprehensive environmental improvement.

Controlling and preventing pollution from livestock and poultry breeding. China will identify the regions where the building of large-scale livestock or poultry farms (communities) is prohibited, and strengthen classified management in different regions. All counties will promote pollution prevention and control in livestock and poultry breeding by waste recycling. Any region with intensive livestock or poultry farms will carry out centralized treatment and recycling of livestock waste. All localities will shut down or move out according to law any livestock farm (communities) and households specialized in breeding in the prohibited areas by the end of 2017. Moreover, China will support transformation and construction of large-scale livestock farms (communities) to meet environmental standards.

Fighting against pollution of agricultural non-point sources. Agricultural structure and layout will be optimized and cleaner production technologies will be promoted in agricultural production, so as to develop agriculture in a resource-saving, environment-friendly, and conservation-oriented manner. A variety of facilities will be built, such as ecological ditches, sewage purification ponds and surface runoff catchment pools, in order to purify farmland drainage and surface runoff. China will carry out an action plan for development of organic agriculture near water. A healthy mode of ecological breeding will be promoted. Soil testing will be conducted before applying fertilizers. Clean production in the planting industry will be promoted through measures such as agricultural film recycling, which enables northeast China to be the first to achieve zero growth of agricultural film in field production in the black soil region. Key regions such as the Beijing-Tianjin-Hebei region, the Yangtze River Delta region and the Pearl River Delta region will conduct research and demonstration projects on prevention and control of key ammonia pollution sources in the planting and breeding industries. China will study and develop post hoc EIA system for pesticide application and work out the Measures on Treatment and Recycling of Pesticide Packaging. China plans to achieve zero growth in chemical fertilizers and pesticides by 2020, with fertilizer utilization rate increased to 40% and recovery rate of agricultural film to 80%. The Beijing-Tianjin-Hebei region, the Yangtze River Delta region and the Pearl River Delta region will meet the above objectives by 2019.

Comprehensively utilizing the straw and putting a ban on straw burning. China will establish an implementation and supervision mechanism on comprehensively utilizing the straw and putting a ban on straw burning at all levels, which combines incentive and restrictive measures with incentives as the main tool. China will improve straw collection and storage system and support commercialization of new technologies such as straw substituting wood, fiber raw materials, clean pulping, biomass energy and commercial organic fertilizer to promote comprehensive use of straw. Then ban on straw burning will be enhanced in key regions and key periods and regulation will be strengthened.

Chapter 6 Integrate Pollution Prevention and Control into the Whole Process to Effectively Prevent and Mitigate Environmental Risks

Measures will be taken to improve the capacity of prevention and control of environmental risks, and regulate risks in a routine manner. A whole-process and multi-layer system will be developed to guard against risks in advance, well regulate risks during the event, and deal with risks after the event. With such a system in place, risks from heavy metals, hazardous waste, toxic and hazardous chemicals, and nuclear and radiation will be under strict control. The regulatory system and capacity building of nuclear and radiation safety will be enhanced to effectively control ecological and social environment risks that threaten public health and keep to the safety bottom line. .

Section 1 Improve the systems of risk prevention and control and emergency response

Strengthening risk assessment and prevention and control from source. The risk assessment system of corporate environmental emergencies will be promoted to regulate the risks of environmental emergencies by category and grading and to strengthen the corporate regulation on risks of major environmental emergencies. Moreover, the identification system of hazardous waste will be improved. Pilots for evaluation of wastewater toxicity, and risk assessment of regional environmental emergencies will be conducted among typical areas, industrial parks, and river basins and the assessment results will serve as reference for industrial access, layout and structural adjustment. In addition, the report on cases of environmental risk assessment in typical areas will be published.

Conducting Surveys, monitoring, and risk assessment on environment and health. Measures on environment and health will be developed to establish a mechanism of survey, monitoring, and risk assessment of environment and health and to form a system of supporting policies, standards, and technologies. Surveys on environment and health will be carried out in key regions, river basins, and industries, and a monitoring network of environment and health risks will be primarily established to identify and assess the environment and health risks in key regions, river basins, and industries. List-based management will be conducted for enterprises and pollutants with environment and health risks, and environmental criteria favorable to human health will be worked out and issued.

Better managing the early-warning on environmental risks. Early-warning on heavy air pollution, drinking water source, toxic and hazardous gases, and nuclear safety, etc. will be strengthened. Pilot projects will be conducted to monitor and issue early-warning on biological toxicity of drinking water sources and toxic and hazardous gases in chemical industrial parks, etc.

Better managing the emergency response to environmental emergencies. The environmental emergency management system that is coordinated across national, provincial, municipal, and county levels will be improved. The trans-regional and cross-sectorial coordination mechanism for emergency response to environmental emergencies will be fully implemented. The integrated emergency rescue system will be improved, and the social emergency rescue mechanism will be established. In addition, the mechanism for on-site command and coordination and the system for reporting and disclosing information will be improved. Moreover, the mechanisms of investigation and analysis, environmental impact assessment, and environmental impact and loss evaluation on environmental emergencies will be enhanced.

Strengthening the fundamental capacity in preventing and controlling environmental risks. A network of environmental risk monitoring and early-warning will be established, covering production, transportation, storage, and disposal, and an IT-based monitoring and management system will be developed, under which the hazardous waste is located, inquired, tracked and assessed, and early warning is released if necessary. A system that supports the command and decision-making of environmental emergencies will be established by improving the database of environmental risk sources, sensitive targets, environmental emergency response capacity and environmental emergency plans, etc. Environmental emergency plans for petrochemical and other key industries, as well as for government and sectors will be well managed. In addition, a training base for national environmental emergency rescue will be established to cultivate high-quality management talents and experts in environmental emergency management. Material reserves and information technology application for environmental emergency response will be enhanced for a strengthened capacity of environmental emergency monitoring. Moreover, the equipment for environmental emergency response will be commercialized and clear standards of environmental emergency response capacity will be established.

Section 2 Step up prevention and control of heavy metal pollution

Strengthening the environmental management of key industries. With a goal to optimize the industrial structure, rapid development of production capacity in heavy metals will be strictly controlled and outdated production capacity will be phased out. Stricter standards for local pollutant emission and environmental access will be developed and implemented in areas where heavy metal industries concentrated and develop rapidly and in a large scale. Those enterprises in heavy metal industry that have no possibility in reaching environmental standards or still cannot meet the standards after improvement and rectification will be shut down. An. integrated plan will be developed on pollution control of industrial parks for industries of electroplating, leather, and lead-acid battery etc., to promote clean and up-to-standard development of related industrial parks. As part of the efforts to have a clear track of environmental risks, monitoring of heavy metal emissions from industrial parks and key mining factories, and of heavy metal concentrations around those parks and factories will be intensified, and information about emission, environmental management, and environmental quality of heavy metal enterprises will be released to the public. The investigation on thallium pollution will be organized in such typical industries as metal mining and smelting, and steel production, as well as on typical regions such as Buyi and Miao Autonomous Prefecture in southwest Guizhou Province. A prevention and control work plan for thallium pollution will be developed based on the above investigation. Moreover, more efforts will be made in regulation of environmental protection projects involving heavy metals in imported mineral products.

Enhancing the prevention and control of heavy metal pollution in key regions based on the categories of heavy metal. The key areas where prevention and control of heavy metal pollution is a priority will develop and implement an integrated work plan on prevention and control of heavy metal pollution so as to effectively prevent and control environmental risks and improve regional environmental quality. The region-specific guidance and policy will be offered for differentiated management. Integrated control of environmental problems in watersheds and regions such as the Xiangjiang River Basin will be speeded up. It is planned that about 20 areas will exit from the list of key regions of pollution prevention and control after considerable efforts during the 13th FYP period. Pilot projects on heavy mental prevention and control will be initiated in 16 key areas such as Jingjiang, Jiangsu Province and Pingyang County, Zhejiang Province, and 8 river basins such as the Fujiang River basin in Dayu County, Jiangxi Province, in order to explore a technical and management system for prevention and control of heavy metal pollution and environmental risks at regional and river-basin levels. A coordination mechanism for integrated prevention and control of environmental pollution will be established for the "Manganese Triangle" (Xiushan County in City of Chongqing, Huayuan County in Hunan Province, Songtao County in Guizhou Province with heavy pollution from manganese mining and production processes) to develop a plan for integrated environmental improvement. In addition, the location of environmental monitoring sites will be optimized in key regions to support the development of a national environmental monitoring system for heavy metals by the end of 2018.

Box 5. Pilot Projects for Prevention and Control of Heavy Metal Pollution

1. Regional integrated prevention and control of heavy metal pollution (16 areas) Jingjiang of Taizhou City (integrated prevention and control in electroplating industry), Pingyang County of Wenzhou City (industrial park upgrade and integrated prevention and control), Changxing County of Huzhou Prefecture (integrated prevention and control in lead-acid

battery industry), Jiyuan City (integrated prevention and control and monitoring of heavy metal pollution), Daye of Huangshi City and its surrounding area (prevention and control of pollution from copper smelting and prevention and control of pollution from past production), Zhubu Port of Xiangtan City and its surrounding areas (prevention and control of pollution from past production), Shuikoushan of Hengyang City and its surrounding area (comprehensive pollution control and industry upgrade), Sanshiliuwan of Chenzhou City and its surrounding area (prevention and control of pollution from past production, monitoring and early-warning of environmental risks), realgar ore region in Shimen County of Changde City (prevention and control of arsenic pollution from past production, and of environmental risk), Jinchengjiang District of Hechi (industrial restructuring and pollution from past production), Xiushan County of Chongqing (comprehensive control of electrolytic manganese industry), Xichang of Liangshan Prefecture (prevention and control of pollution of non-ferrous metal industry and remediation of contaminated site), Wanshan District of Tongren City (comprehensive prevention and control of mercury pollution), Gejiu in Honghe Prefecture (industrial restructuring and prevention and control of pollution from past production), Tongguan County in Weinan (comprehensive pollution prevention and control of non-ferrous metal industry), Jinchuan District of Jinchang City (industrial upgrade and prevention and control of pollution from past production).

2. Integrated management of river basins (8 areas)

Fujiang River basin in Dayu County of Ganzhou City (arsenic pollution), Hongnongjian River basin in Lingbao of Sanmenxia (cadmium and mercury pollution), Lihe River-Nanquan River basin in Zhongxiang of Jingmen City (arsenic pollution), Hengshishui River basin in Dabaoshan mining area of Shaoguan City (cadmium pollution), Diaojiang River basin in Nandan County of Hechi City (arsenic and cadmium pollution), Duliujiang River basin in Dushan County of Qiannan Prefecture (antimony), Bijiang River basin in Lanping County Nujiang Prefecture (lead and cadmium pollution) and Yongning River basin in Huixian County of Longnan City (lead and arsenic pollution).

Stepping up the prevention and control of mercury pollution. Any new PVC production projects that employ mercury-involved calcium carbide method will be forbidden. The mercury consumption per unit of products in the PVC industry will be reduced by 50% by 2020 compared to that of 2010. Mercury emission control in such key industries as coal-fired power plants will be enhanced. In addition, construction of new primary mercury mines will be forbidden, and mining of the existent primary mercury will be stopped gradually. Moreover, products such as mercury thermometers and sphygmomanometers will be eliminated.

Section 3 Enhance hazardous waste disposal

Improving hazardous waste disposal in a differentiated way. Provinces, autonomous regions, and municipalities should evaluate the generation, disposal ability of hazardous waste, work out and implement a plan for establishing facilities to dispose hazardous waste in a concentrated way, incorporating the concentrated disposal facilities of hazardous waste into the local plan on public infrastructure. Industrial bases such as petrochemical enterprises will be encouraged to construct their own facilities for hazardous waste utilization and disposal. Enterprises and industrial parks that produce only one type of product in will be encouraged to construct the facilities to collect, pre-process and dispose hazardous waste. Cement kilns will be guided and regulated in disposing hazardous waste in a coordinated way. Evaluation, and prevention and control will be conducted for accumulative environmental risks of centralized disposal facilities for typical hazardous waste, and some facilities with outdated technologies or failing to meet standards will be eliminated, and some other facilities will be upgraded or managed in a standardized way.

Controlling and preventing the environmental risks of hazardous waste. The national hazardous waste directory will be revised on a regular basis, and the national survey of hazardous waste will be conducted. It is planned that the general situation of generation, storage, use, and disposal of hazardous waste in key industries will be obtained by the end of 2020. Criminal activities such as illegal transfer, use, and disposal of hazardous waste will be cracked down on, especially in petrochemical and chemical industries. Regulation on quality and security of imported petrochemical and chemical products will be strengthened, and import of such solid waste as waste oil in the name of crude oil, fuel oil or lubricants will be cracked down on. Standardized management, supervision and examination on hazardous waste will be continued, and special pollution control actions targeting the heavy metal waste that contain chromium, lead, mercury, cadmium, and arsenic as well as fly ash from municipal solid waste incineration, antibiotic residues, and wastes with highly persistent toxicity will be conducted. Measures on recycling waste lead-acid batteries will be developed. Moreover, the environmental requirements for the control of secondary pollution from use and disposal of hazardous waste and for comprehensive utilization process will be identified, and the upper limit on toxic and hazardous substances in comprehensively utilized products will be set to facilitate safe use of hazardous waste.

Promoting the Safe disposal of medical waste. The scope of service provided by centralized medical waste disposal facilities will be expanded, and a regional coordination and emergency response mechanism will be established. Safe disposal of medical waste in rural areas, towns, and remote areas will be promoted based on local conditions. Medical waste incineration facilities will be renovated to meet higher standards. To manage medical waste in a standardized way, criminal activities such as illegal trade of medical waste will be penalized. An exit mechanism for the franchise of medical waste will be established, and the policy of charging the medical waste disposal will be strictly implemented.

Section 4 Build up the capacity to manage chemical risks

Assessing the environmental and health risks of existing chemicals. Preliminary screening and risk assessment on a group of existing chemicals will be conducted, and the accumulation and risks of chemicals in the environment will be assessed. It is planned that the list of chemicals under prioritized control will be published before the end of 2017, to strictly restrict the production, use, and import of high-risk chemicals and gradually eliminate or replace them. In addition, capacity building for assessing environment and health risks of toxic and hazardous chemicals will be strengthened.

Reducing the chemicals regulated by international conventions. It is planned that chemicals regulated by the *Stockholm Convention on Persistent Organic Pollutants* including lindane, perfluorooctane sulfonate (PFOS) and its salts, perfluorooctane sulfonyl fluoride and endosulfan will be basically phased out by 2020. Research and development on alternatives of the restricted or prohibited persistent organic pollutants, best available technologies, and on relevant monitoring and testing equipment will be enhanced.

Controlling the pollution of chemicals with environmental hormones. It is planned that the survey on production and usage of chemicals with environmental hormones will be completed by the end of 2017. Also, risks of water sources, farmland and concentrated aquaculture areas will be monitored and assessed, and environmental hormones will be eliminated, restricted, or replaced by alternatives.

Section 5 Strengthen management over nuclear and radiation safety

China as a big country of nuclear technology and nuclear power. It is planned that in the 13th FYP period, nuclear safety regulation system and regulation capacity building will be strengthened, and rule of law for nuclear safety will be accelerated. Moreover, the nuclear safety plan will be implemented, and nuclear safety regulation will be strictly conducted according to laws and regulations to prevent any nuclear accidents with radioactive pollution to the environment.

Improving the safety of nuclear facilities and radioactive sources. Safe operation of nuclear power plants will be continuously improved, supervision on the quality of nuclear power plants under construction will be enhanced to ensure that new nuclear power plants meet the latest international standards. Safety of research reactors and facilities for recycling nuclear fuel will be improved. The license management of nuclear safety equipment will be enhanced to enhance, their quality and reliability. Moreover, the action plan for radioactive sources safety will be implemented.

Better controlling and preventing radioactive pollution. Decommissioning of aged nuclear facilities as well as disposal of radioactive waste will be promoted, the ability to dispose radioactive waste will be strengthened reduce waste to the minimum level. In addition, decommissioning of uranium mining and environmental remediation will be facilitated, and regulation on mining and smelting of uranium ores and associated radioactive ores will be strengthened.

Enhancing the regulation system and capacity building for nuclear and radiation safety. Mechanism for regulating nuclear and radiation safety will be strengthened to include the key nuclear safety technologies into the national key research and development plan. Material reserves and capacity building for nuclear emergency response at national, regional, and provincial levels will be enhanced. A national research and development base for nuclear and radiation safety regulation will be developed. A national platform for monitoring, early warning, and emergency response of nuclear safety will be established to improve the network of nuclear radiation environmental monitoring, and to strengthen the regulation capacity of nuclear and radiation safety at national, provincial, and municipal levels.

Chapter 7 Enhance Ecological Protection and Restoration

Following the idea that "mountains, waters, forests and farmlands form a community of shared life", China will put protection first with natural restoration in dominance, promote ecological protection and restoration of key regions and important ecosystems, and build ecological corridors and biodiversity conservation networks to enhance the stability and service functions of various kinds of ecosystems, and consolidate the ecological-security shields.

Section 1 Safeguard national ecological security

Protecting the national ecological security in a systematic way. Identifying important regions of great importance to national ecological security, China will give priority to ecological conservation and safeguard national ecological security with ecological security shields and major river systems serving as the overall structure, national key ecological functional areas as the underprop, areas where development activities are prohibited as the focal points, and ecological corridors and biodiversity conservation networks the framework.

Developing a national ecological-security shields with "two-shields and three-belts". The ecological-security shields of Qinghai-Tibet Plateau will be developed to promote local ecological development and environmental protection, with priority given to protecting diversified and unique ecosystems. The ecological-security shields of Loess Plateau-Sichuan-Yunnan will be enhanced to prevent and control soil erosion and protection of natural vegetation and ensure ecological security of the middle and lower reaches of the Yangtze River and the Yellow River. The northeast forest belt will be developed as ecological-security shields to protect forest resources and biodiversity and safeguard ecological security of Northeast China Plain. Also, China will develop the north shelterbelts as ecological-security shields, and focus efforts on planting more shelter forests, protecting the grassland and promoting sand-fixation. In the desertificated land that cannot be effectively controlled for the time being, development activities there will be prohibited to uphold ecological security in North China, Northeast China and Northwest China. In addition, China will develop the south hills shelter as ecological-security shields to strengthen vegetation restoration and soil erosion control and ensure ecological security of South China and Southwest China.

Developing a biodiversity conservation network. More efforts will be made to implement the *National Strategy and Action Plan on Biodiversity Conservation*. China will continue to engage in the *United Nations Decade on Biodiversity*, developing and implementing local action plans for biodiversity conservation. Measures will be taken to intensify the management of priority areas for biodiversity conservation, develop a biodiversity conservation network, and improve ex-situ conservation facilities to achieve systematic biodiversity conservation. Moreover, assessment and demonstration on the service value of biodiversity and ecological system will be carried out.

Section 2 Manage and protect key ecological regions

Better protecting and managing the national key ecological functional areas. China will develop a negative list of industrial access to national key ecological functional areas and a directory for industries whose development is restricted and prohibited locally. The transfer payment policy will be improved and the stability of regional ecological function and capacity in providing ecological products will be better assessed. The development of a comprehensive demonstration site for Gansu ecological-security shield will be supported and the Yellow River eco-economic belt will be promoted. China will speed up steps to implement ecological protection and development projects in key ecological functional areas, strengthen ecological supervision on development and construction activities, and protect key wildlife resources to significantly improve service functions of eco-systems in those areas.

Prioritizing the development and management of nature reserves. China will optimize the nature reserve layout, make as the priorities of newly built nature reserves the important rivers and lakes, oceans, grassland ecosystems and aquatic organisms, natural relics, wild plant species with extremely small population, and extremely endangered wild animals and develop nature reserve clusters and small-scale nature reserves to manage nature reserves in a systematic and targeted way, with better application of information technology. China will establish a 3-dimensional dynamic monitoring system which unifies information collected from land and space will be developed in nature reserves, with the application of remote sensing technology. National nature reserves will be monitored twice a year and provincial nature reserves once a year. A special inspection on law enforcement related to nature reserves will be carried out on a regular basis, illegal activities will be investigated and severely dealt with, holding those responsible for the damage accountable. In addition, comprehensive scientific exploration, basic investigation and management assessment of nature reserves will be strengthened. China will verify and settle the boundary of nature reserves, identify land-related rights and regulate land use, and gradually relocate the residents living in core zones and buffer zones of nature reserves. It is planned that by 2020, the total area of nature reserves will maintain about 15% of total land area, covering over 90% of wildlife species under priority conservation and typical ecosystems.

Integrating a number of national parks. Pilots on national parks will be well guided and an overall plan for developing national parks will be developed based on the findings of pilot projects. Work will be done to appropriately define the scope of national parks, establish a sound protection system of nature reserves with scientific classification to better protect the authenticity and integrity of natural ecosystems as well as natural and cultural heritages. China will improve the overall coordination of planning, construction and management of various protected areas such as scenic spots, natural and cultural heritages, forest parks, desert parks and geological parks to raise the efficiency of protection and management.

Section 3 Protect important ecosystems

Protecting forest ecosystems. China will improve its natural forest protection system and strengthen the protection and nurture of natural forests. Natural forests will be better managed and protected, and development of protection and management infrastructure will be strengthened to cover the whole protected natural forest. Commercial logging of natural forests will be stopped. China will adhere to its subsidy policy on protecting and nurturing forests as well as developing forests for public interest. China will strictly protect forest resources and manage the use of forestland by grading and category. It is planned that the total forest area will reach 312.3 million hectares by 2020.

Improving the forests quality in a well targeted way. China will adhere to the general principle of giving priority to protection with natural restoration in dominance, and putting equal emphasis on the amount and quality of forest with quality as the priority. Closing hillsides to facilitate afforestation and artificial afforestation will be carried out simultaneously. Land will be closed for afforestation, or planted with forest shrub or grassland based on local conditions. China will enhance forest operation, develop mixed forests, promote restoration of degraded forests and optimize forest composition, structure and functions. It is planned that the mixed forests will take up 45% of total forests, the forest stock per unit of area will reach 95 m^3/ha and forest carbon reserves will reach 9.5 billion ton by 2020.

Protecting grassland ecosystems. China will improve its contracting system for grassland, adhere to the basic state policy on grassland protection and implement the systems such as grass-livestock balance, grazing prohibition and resting and rotational grazing by region. Measures will be taken to strictly regulate the use of grassland , develop teams for forest management and protection, and crack down any infringements damaging grassland such as illegal requisition or occupation, reclamation as well as unregulated and over-exploitation of wild plants. Surveys on grassland resources will be conducted to support the development of a monitoring and early-warning system for grassland production and ecology. In addition, the degraded, salinized or desertificated grassland will be improved and rats, insects and weeds hazards will be controlled or prevented. It is planned that about 30 million hectares of degraded, salinized or desertificated grassland will be treated by 2020.

Protecting wetland ecosystems. China will conduct trial projects on eco-compensation for wetland ecological benefits and conversion of farmland to wetland. China will carry out wetland conservation and restoration projects in international and national important wetlands, wetland nature reserves and national wetland parks; gradually restore ecological functions and expand the total area of wetlands. Moreover, China will also improve its capacity in protection and management of wetland.

Section 4 Enhance ecosystem functions

Carrying out the green land campaign on a national scale. China will carry out a large-scale greening campaign, develop shelter forests for farmland and appropriately distribute urban-rural green land with stable structure and sound functions, and establish green networks along seas, rivers and lakes (reservoirs), railways and highways, borders and islands, so as to promote coordinated greening of mountains, plains, rivers and lakes, cities and villages.

Conducting a new round of "grain for green". The scope and scale of a new round of grain for green will be expanded. The grain for green will be carried out in the farmland with a slope of over 25 degrees, severely desertificated farmland and farmland with a slope of 15~25 degree located at important water sources. China will implement the national plan for grain for green project, steadily expand the project's scope, transform husbandry production mode, and construct grassland protection infrastructure to protect and improve the ecology of natural grassland.

Developing a shelter forest system. China will strengthen the development of shelter forest systems in such regions as the Three-North (North China, Northeast China and Northwest China) areas, the Yangtze River basin and Pearl River basin, Taihang Mountains and coastal areas. Combining arbors, shrubs and grass, the "Three-North" Shelter forest will be developed on a large scale and in a coordinated way that highlights priority areas. In the Yangtze River Basin, the priorities are restoration of degraded forests, improvement of forest quality and development of shelter forest in Dongting Lake, Panyang Lake and Dangjiangkou Reservoir. In Pearl River Basin, degraded forests will be restored. The stand structure of the Taihang Mountains will be optimized. In coastal areas, measures will be taken to develop the coastal backbone forest belt and sea-wave dissipation forests, and improve coastal shelter forest and disaster prevention and relief system. In major grain producing areas, a network of farmland and forests will be developed, and greening of villages and towns will be strengthened, to improve the comprehensive functions of shelter forests in the farmland on plain. .

Developing reserve forests. The provinces and other areas in southern part of China with favorable light, heat, soil and water conditions will attract private capital to investment, operation and management of reserve forests to facilitate the development of reserve forests. Key state-owned forest farms in the Northeast China and Inner Mongolia will cultivate artificial forests in an intensive way take measures such as intensive cultivation of artificial forests, improve, nurture and replant existing forests, and develop reserve forest bases dominated by timber forests and precious tree species. It is planned that there will be 14 million ha. of reserve forests with an annual increase of over 95 million m^3 timber supply by 2020.

Fostering new greening mechanisms. To promote nation-wide participation in greening, China will encourage family-run forest farms, specialized cooperatives organizations in forestry, enterprises, social organizations and individuals to engage in large-scale and specialized afforestation. China will give play to the leading role of state-owned forest fields and farms in advancing national greening conduct diversified cooperative afforestation and forest conservation and cultivation, and encourage state-owned forest farms to bear the main responsibility for regional afforestation and ecological restoration. China will make innovation in property rights and encourage local governments to explore the policy of turning commercial forests into forests for public interest by means of redemption and replacement in ecologically important regions.

Section 5 Restore ecologically degraded areas

Comprehensively dealing with water and soil erosion. The development of water and soil conservation projects will be strengthened in key regions such as the middle and upper reaches of the Yangtze River and the Yellow River, Karst regions in Southwest China and black soil regions Northeast China. Measures will be taken to enhance ditch reinforcement and tableland conservation in the gully areas of Loess Plateau, deal with gully erosion in the black soil region, expedite the control of slope collapse in hilly areas in South China, and actively develop small ecologically clean watersheds.

Facilitating the control of desertification and stony desertification. China will expedite the implementation of the national plan for desertification control, make more efforts in the control and fixation of sand in such areas as major sand source areas, sand mouth, pathways of sand movement and desertification-prone areas, strengthen sand control and prevention along the "Belt and Road", develop the protection zones in desertificated areas where development activities are prohibited and strengthen comprehensive demonstration sites on desertification prevention and control. The Phase II of the Beijing-Tianjin Sand Source Control Project will be continued to further curb sand hazards. With priority given to stony desertification areas in Yunnan, Guangxi and Guizhou and karst regions of the Yangtze River and Pearl River basins, stony desertification will be controlled or prevented in a comprehensive manner. It is planned that China will build ten one-million-mu bases, a hundred 100,000-mu bases and a thousand 10,000-mu bases for desertification prevention and control by 2020.

Better protecting the ecological environment and restoring the ecology of mines. EIA on development of mineral resources will be implemented to develop green mines. China will make more efforts to restore vegetation in mines for a better geological environment, and carry out special campaign targeting tailings ponds with environmental hazards or less than 1 km away from residents or important facilities, and restore the ecological environment of mines with environmental pollution due to the past production. Moreover, Technologies such as tailings backfill will be employed to develop a number of mines with no or only a small amount of tailings and promote restoration and reuse of the abandoned zones in mining areas.

Section 6 Expand ecological products supply

Developing green industries. China will reinforce the development of forestry resource bases, and move fast to restructure and upgrade industries to promote high-end, brand-oriented, specialized and customized production, in a view to meet public demand for quality green products. China will develop some influential demonstration bases for flowers and trees and some demonstration sites in industries able to generate economic benefits, such as forestry for grain and oil, and for non-timber products, forest-based economy, forest biotechnology, sand, and domestication, breeding and utilization of wild animals. Development and improvement of industries such as forest tourism and recreation, wetland vacation, desert exploration and wildlife watching will be expedited, and technical transformation and innovation of forest product and forestry equipment manufacturing industries will be promoted. A number of competitive and distinctive industrial clusters and demonstration parks will be developed. A market monitoring and early warning system will be established for green industries and national key forest products.

Building ecology-related public service network. China will step up efforts to develop public service facilities in nature reserves and in areas designed for people to experience the nature. Quality ecological services and products covering ecological education, recreation and leisure, healthcare and elderly care will be developed. Public service facilities such as labeling system on ecological conservation, greenway network, sanitation and safety facilities shall be put in place at an early date. High-end eco-tourism packages will be developed including forests, wetlands, deserts, wildlife habitat, flowers and trees. Public camps and ecological inns will be built to improve the quality of products and services related to nature experiences.

Enhancing protection and management of scenic spots and world heritages. China will carry out census on scenic spots to make informed decisions when developing and applying for world natural heritages and world natural and cultural heritages. Management of scenic spots and world heritages will be enhanced by conducting dynamic remote sensing for monitoring and strictly controlling ways and intensity of exploiting the tourism resources. Moreover, China will increase its input into facilities for protecting and utilizing scenic spots.

Protecting and restoring urban natural ecosystems. China will increase urban biodiversity by better protecting green areas and properly managing urban greenways. Layout of urban green areas will be optimized, creating greenways and corridors to develop an integrated urban ecological network composed of forests, waters and farmland. Ecological space, such as green areas and waters, will be expanded. Urban green areas will be properly planned, promoting three-dimensional green space and plants on roofs. China will take measures to restore urban mountains, waters, waste areas and green areas, and carry out urban ecological restoration demonstration projects combining natural and human-induced restoration measures. Afforestation in outskirts and urban agglomeration will be promoted by implementing "reforestation of construction land" and growing urban forests. The green coverage rate of built-up urban areas will be raised and transformation of old parks will be accelerated to improve their ecological functions. Ecological afforestation will be promoted by planting native tree species with appropriate mix of trees, shrubs and grass to achieve natural growth. Heritage trees will be protected, and big natural trees are prohibited from being transplanted into cities. China will develop forest cities, garden cities and forest towns. The per capita urban green area is expected to reach 14.6 m^2 and the green coverage of urban built-up areas 38.9% by 2020.

Section 7 Conserve biodiversity

Conducting biodiversity background survey and observation. China will carry out key biodiversity conservation projects. Ecosystems, species, genetic resources and associated traditional knowledge in priority regions for biodiversity protection will be investigated and evaluated in order to establish a national biodiversity database and information platform. It's planned by 2020 there would be a clear understanding of the biodiversity situation based on background survey in the said regions. Biodiversity conservation system will be improved by establishing comprehensive observation stations and sampling areas. There will be observation, monitoring, evaluation and early warning on important biological communities and ecosystems, wildlife under national priority protection as well as their habitats on a regular basis.

Rescuing and protecting endangered wildlife. China will protect, restore and expand the habitats of rare and endangered wildlife species and in situ conservation areas and sites; prioritize the implementation of key projects on the protection of wild animals under national priority protection and wild plants with extremely small populations; develop technologies on breeding, restoration and conservation of endangered species; and rescue and protect those rare and endangered species, facilitate their breeding and release them to the wild. In the Yangtze River Economic Belt as well as other key watersheds, there will be pilot projects on releasing animals to the wild. There will be steps to reintroduce rare and endangered wildlife species in a scientific manner, improve the national wildlife rescue network, and build wild animal rescue and breeding centers based on good planning, and to establish centers to breed rare and endangered plant species like the orchids.

There will be intensified efforts to supervise how wildlife and their products are used through certification and labeling. *The List of Endangered and Protected Species of China* will be updated.

Conserving biological and genetic resources. China will establish a system that enables access to biogenetic resources and relevant knowledge and shares the benefits with concerning parties. It will also regulate how biogenetic resources are collected, conserved, exchanged, studied, developed and utilized, and strengthen the protection of traditional knowledge on biogenetic resources. Biogenetic resources will be carefully evaluated, including steps to sort, test, cultivate and examine the character of biological resources, and to screen and select genes of species with favorable characters. Wildlife genes will be protected by constructing conservation bases and gene banks. The existing biogenetic resource bank in southwest China will be updated and a new one will be built in the central and eastern region to collect and conserve unique, rare, endangered, and valuable biogenetic resources. Medium to long-term repositories and nursery garden will be built for conserving biological resources including medicinal plants and crop germplasm, wild flowers and forests. Planning and building of botanical gardens, zoos, and wildlife breeding centers will be completed.

Enhancing regulation of wildlife import and export. China will strengthen the management of import and export of biogenetic resources, wildlife and wildlife products; establish a working mechanism for information sharing and joint prevention and control, and develop and improve electronic network on importing and exporting information. There would be tough measures against illegal trade of wildlife products such as ivory. A platform for intelligence and information sharing and analysis as well as cooperation on combating illegal trade will be developed. Unique, rare and endangered wildlife germplasm resources will be well protected from loss.

Guarding against biosafety risks. China will strengthen the protection of wild animals and plants from epidemic diseases, including efforts to set in place a dynamic monitoring and early-warning system on national ecological safety so as to thoroughly investigate and assess ecological risks on a regular basis; to regulate the ecological release of genetically modified organisms through risk assessment and follow-up monitoring when releasing them into the environment; and to establish a border biosafety network by improving border biosafety inspection mechanism and strictly controlling the introduction of alien species. There will be tough measures against invasion of alien hazardous species, including conducting census, monitoring and ecological impact assessment of invasion of alien species, as well as controlling and eradicating such species that cause significant ecological impact.

Chapter 8 Modernize Governance System and Enhance its Capacity by Accelerating Institutional Innovation

China will take a holistic approach to push forward a sound environmental governance system, which consists of a range of mechanisms involving the government, business, and civil society. To be specific, to ensure the local environmental authorities fulfill their due role, there will be environmental inspection tours from central environmental authorities, a system of creating a balance sheet for natural resource assets, auditing outgoing officials' manage of natural resource assets, as well as an accountability system for damaging the environment. To ensure the business fully comply with the law and take on its due responsibility, there will be steps to strengthen environmental justice, and to introduce systems on pollution discharge permit system and on damage compensation, etc. To engage the civil society into the governance system, market-based measures will be introduced such as environmental information disclosure, public interest litigation, and green finance.

Section 1 Put in place a sound legal framework

Putting in place a sound legal framework. It's required to revise laws and regulations concerning resources and environmental protection, and further refine environmental systems on preventing and controlling water, acoustic, and soil contamination, as well as on ecological compensation and nature reserves in due course.

Strengthening environmental law enforcement supervision. The supervision mechanism for environmental law enforcement will be improved through enforcing environmental laws in joint efforts, through regional cooperation, and in collaboration with different departments. Law enforcement will be under strict supervision, and those who fail to enforce a law will be held accountable. It's necessary to clearly define the responsibilities of different divisions within the environmental law enforcement agencies such as administrative investigation, administrative punishment, and administrative enforcement, so that law enforcement and supervision forces in different fields and departments at different levels will be further integrated, and these forces will be brought to grass-root levels.

Reinforcing Environmental justice. There must be a strengthened link connecting administrative law enforcement with environmental justice, especially concerning consistency of procedures, in case transfer, and in application for compulsory enforcement. The environmental protection authority must enhance its communication and coordination with and the authorities of public security, people's procuratorates and people's courts. Moreover, the system for trial of environmental cases shall be improved by revising the judicial interpretations of relevant laws and regulations in collaboration with the judicial authorities.

Section 2 Improve market mechanism

Carrying forward the emission trading system. China will set in place a sound system on allocating and trading emission permits, including introducing a paid use system for emission permits, implementing pilot projects on paid use and trading of such permits, and creating platforms for such purpose. It is encouraged to obtain the emission permits by trading when establishing new projects to ensure the total emissions are not increased in this region. There will also be a system to manage the budget of energy cost, as well as pilot projects on paid use and trading in this regard.

Financial and tax policies playing a guiding role. Environmental tax will take effect. Resource tax shall be reformed and made applicable to all natural ecological space. There will be more supporting policies in granting preferential tax on acts conducive to environmental protection, ecological conservation, and development and application of new energies. In addition, policies will be introduced to prepay the expense of decommissioning key centralized disposal facilities and sites of hazardous waste.

Pricing reforms on resources and the environment. Pricing mechanism must be further reformed to make prices of resources and the environment fully reflect the market supply and demand, resource scarcity, environmental damage costs, and remediation benefits, etc. There will be steps to adjust charges of sewage treatment fees and water resources fees, charge waste disposal fees on a broader basis, and refine the pricing mechanism for reclaimed water. There will be doubled efforts to refine the pricing policies for coal-fired power plants that are environment-friendly, and implement differentiated power charging policies on industries high in energy and water consumption and in pollution.

Encouraging businesses to engage in environmental management. There will be efforts to explore a development mode that combines environmental management projects with business projects, and to refine the return on investment (ROI) mechanism for private investment in environmental management. It's also planned to continue pilot projects on environmental services, innovate in pollution control models and management models such as regional integration in management environmental issues, "Internet + " for environmental protection, and network of things on environment. Various funds and capitals shall be channeled into the field of environmental protection. Regulations that impede a unified national market and fair competition will be repealed, and a credit system for market-based environmental governance will be set in place to better regulate market behaviors. In addition, it is encouraged to implement efficiency-based environmental management payments and environmental performance contracting services.

Building a green finance system. China will establish a green rating system, as well as a system to account environmental cost and assess its impact that is on a public nature. There must be clear and detailed statement on a lender's due diligence and exoneration, and his or her environmental protection responsibility, and financial institutions are encouraged to release more green credits. Mandatory environmental pollution liability insurance system will be established in those sectors with high environmental risks. It's planned to study and develop green stock index and relevant investment products. Banks and enterprises are also encouraged to issue green bonds and turn more green credit assets into securities. Moreover, more efforts will be taken to compensate risks, and to support guarantee business and mortgage loans on emission permits, charging permits, and service purchase agreement. It's also encouraged to develop green development funds of various kinds that are operating under market mechanism.

Pushing ahead a diversified eco-compensation mechanism. China will make more transfer payments to areas with key ecological functions, and appropriately raise compensation standards in favor of ecologically sensitive and vulnerable areas and river basins. There will be incentive mechanism that links ecological protection performance with fund allocation, and diversified tools including fund, policy, industry, and technology that are thought complementary with each other will be employed to bring each and every factor into full play. Eco-compensation will be promoted to cover a wider range of areas, in an effort to achieve a full coverage of all key ecosystems of forest, grassland, wetland, desert, rivers, ocean, and farmland, as well as areas that are prohibited from development or with key ecological functions. There will be central funds supporting the local authorities in implementing a trans-provincial compensation mechanism by which the ecologically benefited areas and the areas protecting the environment, as well as the upstream areas and downstream areas, are able to compensate each other for ecological or environmental damage and ecological conservation efforts. Pilot projects will be launched on ecological compensation targeting key rivers basins of the Yangtze River and the Yellow River. There will be efforts to protect drinking water source areas of the Middle Route of South-North Water Diversion Project and pilot projects on ecological compensation for water environment of the Xin'an River. Likewise, there will also be pilot projects on trans-regional ecological compensation in regions of the Beijing-Tianjin-Hebei Water Conservation Zone, Jiuzhou River in Guangxi Autonomous Region and Guangdong Province, Dingjiang River - Hanjiang River in Fujian Province and Guangdong Province, Dongjiang River in Jiangxi Province and Guangdong Province, and Xijiang River in Guangdong Province, Yunnan Province, Guizhou Province and Guangxi Autonomous Region. It is planned by 2017 an ecological compensation mechanism shall be in place in the Beijing-Tianjin-Hebei region, by which Beijing and Tianjin shall be able to support the Hebei province in conserving the ecosystems and protecting the environment.

Section 3 Ensure localities fulfill their due responsibilities

Ensuring the governments performing environmental protection duties. China will establish a sound, environmental protection responsibility system featuring clearly defined power and responsibilities and properly allocated tasks. Supervision and inspection will be further strengthened to ensure leaders of local CPC committee and local government are responsible for both economic development and ecological progress, and share the same responsibility in this regard. Provincial governments take the primary responsibility in protecting the environment and resources within its administrative region, including the responsibilities for protecting the river basins that are located within its administrative region, and for coordinating and improving basic environmental service to make it equitable and accessible to all. Municipal governments are responsible for coordination and integrated management of the environment, and the county-level governments primarily for implementation.

Advancing institutional reform on environmental protection. China will pilot the reforms for conducting direct oversight by provincial-level environmental protection agencies over the environmental monitoring and inspection work of environmental protection agencies below the provincial level, and strengthen supervision and inspection on the performance of local governments and relevant environmental authorities in protecting the environment. There will be management models on trans-regional and inter-watershed prevention and control of environmental pollution, as well as urban-rural joint governance of environmental protection. An environmental management system that strictly restricts the emission and discharge of all pollutants will be established and improved.

Promoting strategic environmental assessment and environmental impact assessment (EIA) on development plans. By the time when strategic environmental assessment (SEA) are completed in the Beijing-Tianjin-Hebei region, the YRD, the PRD, the Yangtze River Economic Belt, and the areas involved in the "BRI", SEA will then gradually extend to others regions at provincial and municipal levels. There will be pilot projects to assess environmental impacts generated from some important policies. Major development and construction activities shall subject to environmental impact assessment which will form important basis for the development plan to be formed, approval, and implemented. There will be environmental impact assessment on urban development plan and plan on establishing new development zones to better protect the ecological space, and the consultation mechanism for such assessment will be further refined. With industrial parks as the major focus, there will be efforts to develop and better manage the ecological and environmental access list, and to reform how EIA on construction projects within industrial parks is approved and managed. There will be a strengthened link connecting EIA on construction projects and on its planning. There will be an information network on examining and approving EIA between environmental protection authorities at national, provincial, municipal and county levels. Local governments and relevant authorities must, based on SEA and EIA results, translate such requirements as space control, total amount control of pollutants, and environmental access criteria into hard constraints to regulate regional development. It's required to enhance supervision on local governments and relevant authorities when they are assessing environmental impacts of development plans and hold those who fail in their duties accountable.

Creating balance sheets for natural resource assets. It's planned to create balance sheets for natural resource assets. There will be steps to establish accounts for accounting natural resources in physical terms, to put in place an ecological valuation system, and to assess and account ecological assets. A system on auditing outgoing officials' management of natural resource assets will be in place to push local leaders to manage natural resource assets in a responsible manner. Building on those pilot projects on developing balance sheets of natural resource assets, it's expected to formulate step by step a sound mechanism for balance sheets of natural resource assets, and come up with major approaches for valuating and accounting natural resource assets at a national level.

Establishing monitoring and early-warning mechanism for resource and environmental carrying capacity. China will study and develop a monitoring and assessment system, an indicator systems for early-warning, and relevant technical methods. In terms of resource and environmental carrying capacity, there will be monitoring and early-warning when its capacity is close to the upper limit, as well as study and analysis on it. For regions where resource and environmental carrying capacity is approaching or exceeding the limit, warnings will be issued and differentiated restrictive measures will be adopted to prevent it from getting worth. Provinces, autonomous regions and municipalities shall evaluate the resource and environment carrying capacity at city and county levels, and the regions exceeding its carrying capacity should adjust their development plans and industrial structures.

Assessing and examining the performance in promoting ecological civilization. China will implement the *Measures on Examining and Assessing Ecological Progress*. It will put in place a target system, examination methods, as well as "carrots and sticks" mechanism that meet the needs of ecological civilization, incorporate resource consumption, environmental damage and ecological benefits into the economic and social development evaluation system for local governments at all levels, and take differentiated approaches when assessing and examining the environmental performance of different localities based on different functions.

Continuing with environmental protection inspection. China will proceed with environmental protection inspection as part of the efforts to urge local governments in taking on environmental protection responsibilities. The inspection will focus on regions and watersheds where environment is degrading to see whether this trend is reversed. It will target local party committees and governments, and relevant environmental authorities who do not take actions or take wrong actions, and examine how party officials and government officials are sharing the same responsibilities, and are fulfilling their responsibilities in both economic development and ecological progress, as well as whether or not hold those who fail in their duties accountable. All of these will serve to promote local ecological progress and environmental protection in a bid to advance green development.

Establishing a lifelong accountability system for ecological and environmental damages. A lifelong accountability investigation mechanism for major decisions will be established. Any leader responsible for serious damage to the ecological environment and resources shall not be promoted or transferred to important positions, and where a crime is constituted, the leader responsible for it shall be investigated for violating the criminal law. Outgoing officials must subject to natural resource assets auditing. A system of lifelong accountability will be put into effect for major ecological and environmental damage which becomes apparent after an official has left office and for which he or she is found liable.

Section 4 Strengthen regulation of enterprises

Establishing a business emission permit system covering all stationary pollution sources. China will fully implement a pollutants discharge permit system, aiming at improving environmental quality and guarding against environmental risks. This system includes types of pollutants, its concentration, total amount of discharge, and where to discharge, etc. into the scope of permit management system. Enterprises will operate and discharge pollutants according to the requirements in the system. The major responsibility system for pollution control will be improved, where environmental authorities shall be responsible for regulating the enterprises in pollution discharge and enforce relevant laws according to requirements as specified in the permits. It is planned that by the end of 2017 such permits have been approved and issued to the enterprises in key industries and industries with excess capacity, and a national information platform for pollution discharge permit management will be set up. And by 2020, discharge permits have been issued to enterprises specified in the management directory.

Ensuring enterprises fulfill environmental responsibility through "sticks and carrots". An environmental credit rating system for blacklisting enterprises with environmental violations of pollutants discharge will be set up, and corporate environmental violations will be recorded in the social credit files and open to the public. A mandatory environmental information disclosure mechanism will be established, and enterprises failing to fulfill the obligation will be punished according to laws and regulations. A "front runner" system for Energy efficiency and environmental protection will be conducted, and enterprises will be encouraged to achieve higher environmental standards and targets through incentives such as preferential taxation and honor recognition. It is planned to establish a rating system on grading enterprises by their environmental credits, and integrate enterprise environmental credit information into national credit information sharing platform, and to establish a mechanism rewarding the enterprises with good credit and punishing those with poor credit by 2020.

Implementing and refining the assessment and compensation system on ecological environment damage. Standard procedures will be introduced to identify and assess ecological environment damages, and relevant techniques and methods for such purpose will be improved. It is planned to complete pilot projects on the reform of ecological environment damage compensation system by the end of 2017. And starting from 2018, an ecological environment damage compensation system shall be put on trial, in an effort to set up a national environment compensation system by 2020.

Section 5 Mobilize public support

Raising public awareness of environmental protection. China will strengthen publicity and education on environmental protection, organize environmental protection activities for public good, and develop relevant cultural products, with a view to raising public awareness of environmental protection in a thorough manner. Local governments at all levels, education authorities, and news media shall fulfill their responsibilities for publicity and communications according to law and regulations, and promote and practice environmental protection and ecological civilization as an important component of the core socialist values, and implement a national action plan to involve the public into environmental protection communication and education. People will foster a social norm that advocates hardworking, frugality, green and low-carbon development while resisting excessive consumption, extravagance and waste of resources and energy. It's also encouraged to produce more cultural works concerning ecological conservation and environmental protection, develop more and diversified products for environmental protection publicity, and organize environmental protection activities for public good. A national environmental education platform will be set up to guide a shift towards green, simple and low-carbon lifestyle. All primary, and middle schools, high schools, colleges and universities, vocational schools, and training institutions shall be required to incorporate ecological civilization education into their teaching programs.

Advocating green consumption. China will enhance public awareness of green consumption, and motivate attitude and behavioral change towards environmental protection, in an effort to expedite a transformation towards green consumption that covers all basic life necessities. China will carry out a national action plan for energy conservation, including a tiered pricing mechanism for residential use of water, electricity, and gas, and promotion of water-saving, energy-efficient products and green, environment-friendly furniture and building materials. There will also be an action plan for green buildings, by which the standards and certification system on green building will be further improved, and more buildings shall subject to mandatory law enforcement. It is planned that at least 50% of new buildings in Beijing-Tianjin-Hebei urban areas will meet the green building standards. China will strengthen its green public procurement system, develop a green product list for government procurement and encourage non-governmental organizations and enterprises into green procurement. Green travel shall be promoted through fostering favorable conditions for walking and use of bicycles and improving urban public transit system. It is planned that public transportation will account for 60% of motorized travel in built areas of cities with permanent population of over three million by 2020.

Strengthening information disclosure. China will establish a unified mechanism for releasing environmental monitoring information. Information the ecological environment of air, water and soil must be released to the public, and regulatory authorities are required to disclose information concerning ecological environment, pollutants emitters and EIA results of construction projects. All provinces, autonomous regions, and municipalities are required to establish a unified information disclosure platform as well as a sound feedback mechanism. There will also be a sound environmental protection spokesperson system in place.

Reinforcing public oversight. China will facilitate public participation in the environmental management decision-making by establishing effective channels and appropriate mechanisms, and encourage public oversight of government performance in environmental protection and pollution discharge of enterprises. There a development plan is applied, implemented, and assessed upon conclusion, there must be a communication platform in place for public comments and suggestions so as to safeguard their the right to be informed, to participate, to be heard, and to oversee. There will also be steps to guide the news media, strengthen supervision by public opinions, and make full use of "12369" environmental protection hotline and of WeChat environmental reporting platform. It's planned to study and roll out a system whereby typical environmental cases will be released as examples to guide the public, leverage the role of judiciary authorities in safeguarding the rights of public to file environmental lawsuits, specify legal procedures for environmental public interest litigation, provide more technical support to such litigation and improve relevant systems.

Section 6 Build up governance capacity

Improving environmental monitoring network. Based on a unified planning, distribution of monitoring sites on environment quality shall be further optimized, and a national environmental quality monitoring network will be set in place that covers all essential elements of air, water, soil, noise and radiation with properly distributed sites and fully developed functions, so that the information on ecological environment is able to be collected, integrated and shared through this network. It's aimed that the monitoring sites on air and surface water quality will cover 80% of urban districts and counties, as well as all those urban districts and counties that are densely populated. Monitoring sites on soil quality are also expected to realize a full coverage. China will build up capacity in better forecasting air quality and issuing early warning on air pollution, and tracking and analyzing pollution source. There will be air quality forecast in cities at or above prefecture level, as well as a national monitoring and early earning platform on water quality. There will be intensified efforts to monitor persistent, bio-accumulating and hazardous pollutants in drinking water sources and soil, and strengthen monitoring and early warning of the water quality and radioactivity of urban centralized drinking sources in key river basins. Moreover, China will establish an integrated ecological monitoring system that covers air, ground and space and builds on networked operation of environmental satellites, UAV-based remote sensing and ground ecological monitoring. In addition, it's also planned establish a biodiversity observation network.

Box 6 Development of National Monitoring Network on Ecological Environment

1. Steadily centralize power of monitoring on environmental quality

Currently in China, there are a total of 1,436 automatic urban air quality monitoring stations, 96 regional air quality monitoring stations and 16 air quality background monitoring stations, 2,767 state-controlled surface water monitoring sections, 419 coastal water quality monitoring sites and 300 automatic water quality monitoring stations as well as 40,000 national soil environmental monitoring sites. All of these stations and sites shall be managed and funded by the governments, who shall invited a third party to provide monitoring service, and entrust the local authorities with their daily maintenance, management, and director monitoring, in a bid to connect, share and release to the public all environmental monitoring data.

2. Move fast in setting up an ecological monitoring network

China will establish a remote sensing monitoring system on ecological conservation that covers air, ground, and space. There will be a ground monitoring station on ecological functions, as well as an UAV based remote sensing, to monitor areas with key ecological functions and release it to the public in a unified manner. China will establish a national monitoring platform and a number of relatively fixed ground checkpoints for ecological protection red lines. China will set up a biodiversity observation network and monitor and observe important ecosystems and biological populations on a regular basis. In addition, China will establish a number of monitoring facilities, including 400 automatic monitoring stations for atmospheric radiation, 163 soil radiation monitoring sites, 330 radiation monitoring sites for drinking water sources, as well as 228 forest monitoring stations, 85 wetland monitoring stations, 108 desert monitoring stations and 300 biodiversity monitoring stations.

Building up capacity in environmental regulation and law enforcement. China will push forward grid management in exercising environmental regulation, optimize the allocation of regulatory forces, and extend regulatory services to rural areas. It's planned to refine the system of personnel selection, training and performance evaluation in terms of environmental regulation and law enforcement, to supplement front-line law enforcement teams, provide better and sufficient law enforcement equipment, build up capacity in on-site investigation and evidence-gathering, in an effort to build a professional and competent team on environmental regulation and law enforcement. China will carry out a national program of two-way exchanges of environmental protection personnel between the eastern region and the central and western region, and is going to make the personnel in central and west China more capable of environmental regulation and law enforcement. It is planned that all environmental supervision and law enforcement professionals in environmental institutions at all levels will obtain qualifications after going through relevant trainings, and all county-level environmental regulatory agencies shall be equipped with sufficient equipment that could meet the regulation and enforcement needs by 2020.

Establishing and refining environmental protection information system. China will organize the second national census on pollution sources, develop and update the National List of Basic Pollution Sources (enterprises, institutions, etc.). China will strengthen its statistical work on environment, including small and micro-sized enterprises into the statistical framework, sorting out data on pollutants discharge, and gradually integrating various datasets to and ensuring they reflect the actual conditions. China will establish basic database and information management system for typical ecological zones. It will develop and improve a unified, full-coverage, real-time online environment monitoring and control system; expedite the development of big data platform on environment information; integrate and update data about environmental quality, pollution sources, environmental law enforcement, environmental assessment management, natural ecology, nuclear and radiation; establish a platform for disclosing and sharing information and launch pilot projects on environmental big data. It's planned to push ahead smart environmental management through advanced technologies that highlight automatic and intelligent pollution control technologies so as to develop a system for sharing environmental data, products and services.

Box 7 Strengthen Basic Investigation on Ecological Environment

China will strengthen basic environmental survey, including carrying out the second national census of pollution sources, and imitate investigation into hazardous waste, the performance in protecting centralized drinking water sources and rural centralized drinking water sources, groundwater pollution, environmental hormones, invasive alien species, sediment of rivers and lakes in major regions, and into the progress in practicing green lifestyle. There will also be detailed investigation on soil contamination, comprehensive investigation on biodiversity, as well as baseline investigation on resource and environment of national nature reserves. It's planned to complete the *National Ecosystem Survey and Assessment of China* (2011-2015), complete baseline environmental survey on groundwater, survey and assess ecological risk, public awareness in ecological civilization, and ecological health of the Yangtze River basin, as well as push ahead monitoring, survey and risk assessment of environment and health.

Chapter 9 Implement National Projects on Protecting the Ecological Environment

During the 13th Five-Year Plan period, China is expected to roll out up to 25 key projects to bring emissions from all industrial pollution sources in line with the set standards. A project database will be established accordingly to facilitate project management and performance review. These projects are mainly funded by enterprises and local governments, and partly supported by the central budget.

Box 8 Key Projects on Pollution Control and Environmental Protection

1. Up-to-standard discharge for all industrial pollution sources

China will retrofit all 500,000 t/h coal-fired boilers and upgrade wastewater treatment facilities in industrial parks within a given period of time. In cities at or above prefecture It's also planned to retire coal-fired boilers that have a capacity below 100,000 t/h, and complete desulfurization, denitrification and dust removal for the coal-fired boilers, desulfurization of sintering machines in iron & steel industry, and denitrification in cement industry. In addition, any enterprises failing to live up to the standards in industries of iron & steel, cement, plate glass, papermaking, printing and dyeing, nitrogen fertilizer and sugar must go be retrofitted and upgraded. Moreover, industrial parks must be armed with better wastewater treatment facilities.

2. Coal-to-gas application in key regions with heavy air pollution

China will construct and improve natural gas infrastructure such as gas transportation pipelines, urban gas pipelines, gas storages, and peak-shaving stations in regions of Beijing-Tianjin-Hebei, the Yangtze River

Delta, the Pearl River Delta and Northeast China;. China will promote "coal to gas" projects in key cities to replace coal-fired boilers with total capacity of 189,000 t steam.

3. Ultra-low emission retrofit to coal-fired plants

China will finish the ultra-low emission retrofit to coal-fired generating units with total installed capacity of 420 million kW, implement up-to-standard retrofit to 110 million kW, and phase out before the deadline a total of 20 MW capacity of generating units either with backward capacity or unable to meet the binding standards.

4. Integrated prevention and control of VOCs emission

VOCs emission needs to be controlled and treated in all petrochemical enterprises, and comprehensive approach will be adopted to control its emission in organic industrial parks, pharmaceutical industrial parks and coal chemical industrial bases, as well as in industrial coating and package printing industries. Oily and gas vapor recovery as well as integrated management shall be applied to oil stations, oil tankers and bulk gasoline terminals.

5. Protection of good water bodies and groundwater environment

China will take a series of strict steps to protect major headwaters and another 378 rivers, lakes and reservoirs with water quality better than Grade III, and to control and tackle pollutions in the sewage outlets to the said rivers, lakes and reservoirs. Important drinking water sources must live up to national standards, and meanwhile more reserve water sources shall be developed. In addition, efforts will also be made to conserve water source, restore the damaged ecosystems, and develop biological buffers. It's also required to shut down and refill the retired mines, boreholes and water intake wells that were once excavated, and to launch pilot projects to restore groundwater n in key regions of Beijing, Tianjin, Hebei Province and Shanxi Province, as part of the efforts to strengthen groundwater protection.

6. Water environmental governance in key watersheds and sea waters

In view of prominent water problems in seven major river basins and coastal waters, China identified 580 priority control units to implement integrated management and protection measures on rivers. There will be integrated plans to control point and non-point pollution sources, and restore ecosystems in rivers and lakes, and differentiated approach to manage different categories of watershed and improve its environment. All of these require intensified efforts to effectively improve the water quality in key watersheds and sea waters. In key lakes and reservoirs including the Taihu Lake, Dongting Lake, Dianchi Lake, Chaohu Lake, Poyang Lake, Baiyangdian Lake, Ulansuhai Lake, Hulun Lake and Ebinur Lake, integrated measures will be taken to prevent and control its water pollution, while for the middle and lower reaches of the Yangtze River and the Pearl River, steps to control the pollution from their internal sources like the sediment will be adopted.

7. Full coverage of urban sewage treatment facilities

Focusing on prevention and control of urban black and putrid waters as well as 343 control units in need of water quality improvement, China will strengthen sewage collection and treatment and double its efforts to deal with waters of heavy pollution. China will reinforce the construction of sewage treatment facilities in cities, counties and key towns, speed up the construction of pipelines and upgrade sewage treatment plants so that they will meet Grade A discharge limits. In addition, China will promote reuse of reclaimed water, and enhance environmentally sound treatment and disposal of wastewater.

8. Comprehensive renovation of rural environment

There will be special campaigns targeting rural solid waste. Steps shall be taken to address pollution and enhance management over 130,000 administrative villages in China, including carrying out demonstration project on recycling of agricultural waste, building facilities to collect, treat and recycle sewage and solid waste, and treating sewage in a progressive manner. With all these efforts, it's hoped up to 90% of solid waste in the said villages would be properly treated. Moreover, it's also planned to treat waste discharged from large-scale livestock or poultry farms (communities) and turn those wastes into resources by adopting integrated measures and equipping at least 75% of them with solid waste storage and sewage treatment facilities.

9. Prevention and remediation of soil contamination

There would be detailed investigation on soil contamination and efforts to develop a risk identification system on soil environmental quality. It's planned to launch pilot projects to address contamination in 100 plots of agricultural land and 100 plots of construction land, and develop 6 pioneer areas where comprehensive measures must be adopted to prevent and control soil contamination. A total of 10 million mu contaminated arable land will be remedied, and another 40 million would subject to risk control and prevention measures. For the contaminated land which previously accommodated the chemical enterprises that are re-located, remediation plan must be provided based on detailed investigation, and pilot projects would be carried out. For the contaminated land temporarily not scheduled for development, there must be steps to prevent and control its risks. Moreover, there would be steps to deal with pollution and improve environment in mine dumps. Projects must be carried out to remedy heavy metal-contaminated sites, watercourses and solid waste storage sites with high environmental risks, as well as 31 legacy sites with chromium pollution.

10. Prevention and control of environmental risks in key areas

There would be concrete measures to treat and dispose fly ash resulted from municipal solid waste incineration, and to build a regional recycling network for waste lead-acid batteries and lithium batteries. China will strengthen capacity building in assessing environmental and health risk of toxic and hazardous chemicals, set up a basic database of hazardous chemical properties, and establish the National Center for Chemical Computational Toxicology and National Chemicals Testing Laboratory. There would be 50 demonstration projects on environmental risk control that runs through the whole process of business activities in large chemical industrial parks, centralized drinking water sources, and other risk-vulnerable areas. It's expected to establish a national training base for environmental emergency relief with such major functions as personnel training, materials reserves, exhibition, emergency relief, logistics, as well as research and development. A number of supporting systems would be put in place accordingly, including an emergency drill system, a simulation and training site and an online training platform for environmental emergency response. Several platforms are also underway including a big data platform on national ecosystems and environment, a monitoring and early-warning platform on national water quality, a monitoring and control platform on national environmental protection, as well as a monitoring and early-warning platform on air quality in key regions. There will also be a series of atmospheric environmental monitoring satellites and follow-up satellites for network operation. In addition, to facilitate administration and law enforcement, more instruments and equipment would be provided to facilitate investigation and evidence collection for the relevant authorities at city and county levels in central and western region and county-level law enforcement agencies in eastern underdeveloped areas.

11. Capacity building on nuclear and radiation safety

China will establish a R&D center to enhance nuclear and radiation safety, move fast to decommission nuclear facilities that have been operating for years, and dispose radioactive waste left unsolved in the past. It's also planned to build 5 disposal sites for medium- and low-level radioactive waste and an underground laboratory for disposal of high-level radioactive waste as well as a real-time monitoring system for high-risk radioactive sources to ensure all waste or used radioactive source are safely collected and stored. Moreover, a national workforce for emergency response to nuclear accidents shall also be established and reinforced.

Chapter 9   Implement National Projects on Protecting the Ecological Environment   185

Box 9 Ecological conservation and restoration projects for mountains, waters, forests and farmland

1. Protection and restoration of national ecological-security shields

Intensified efforts shall be made to restore ecosystems in key areas of great significance to national ecological security, including the Tibetan Plateau, Loess Plateau, Yunnan-Guizhou Plateau, Qinba Mountains, Qilian Mountains, Greater and Lesser Khingan Ranges and Changbai Mountains, Nanling mountainous areas, Beijing-Tianjin-Hebei water conservation areas, Inner Mongolia Plateau, Hexi Corridor, Tarim River Basin, and Guangxi-Guizhou-Yunnan karst regions.

2. National land greening

It's planned to launch and scale up afforestation campaigns in China, which includes steps to develop contiguous forests, push forward shelter forest program in the "Three-North"(the North, the Northeast and the Northwest China), coastal areas as well as the Yangtze River and Pearl River basins, and to develop reserve forests and timber forest bases. It also calls for efforts to restore the degraded shelter forests, and develop green ecological space and ecological corridors connecting various ecological spaces. Shelter forests and afforestation campaigns will also cover the farmland and the Taihang Mountains, and there would be demonstration projects on afforestation in saline areas and dry and hot valleys. Ecosystems in damaged mountains will be restored.

3. Comprehensive land improvement

There would be comprehensive steps to improve environmental conditions in key river basins, coastal zones and islands, and intensified efforts to restore the geological environment and ecosystems in areas with intensive mining activities. Efforts would also go to reclaiming the damaged land as well as industrial and mining wasteland, and restoring land and mountains damaged by natural disasters and large-scale construction projects. It's planned to treat pollution and improve environmental conditions along the Beijing-Hangzhou Grand Canal and the abandoned watercourse of the Yellow River in Ming and Qing Dynasty, as well as to improve the land environment in border areas through proper development, sound protection and pollution prevention and control.

4. Natural forest conservation

The conservation plan must not be limited to protecting the existing natural forest, but also including such potential natural forest areas as young afforested land that is refrained from economic activities, woodland and shrub land. Forest land unable to regenerate itself must be restored through afforestation.

5. New round of "grain for green" project

"Grain for green" project is expected to be further promoted and applied to a wider range of areas, covering not only arable land that has a slope of above 25 degrees or is severely desertified, but also important water sources with a slope of 15-25 degrees. A larger area of grassland-converted grazing land is expected to return to grassland, and more central fund would be allocated to ensure a sound progress. Up to 10 million hectare of grassland would be fenced from animal grazing, and 2.67 million hectare of degraded grasslands would be restored. The plan also requires 330,000 hectare of grasslands to be planted for animal grazing and 300,000 feeding stalls (storage huts and silage kilns) constructed, and also calls for efforts to improve environment for 330,000 hectare of grassland in the Karst region, restore 70,000 hectare of severely degraded grassland which is known in China as "heitutan", a kind of bare land naked of any vegetation, and clear toxic herbage for 120,000 hectare of degraded grassland.

6. Comprehensive prevention and control of desertification and water and soil erosion

Comprehensive measures would be taken to prevent and control water and soil erosion in key regions such as shelterbelts in North China, Loess Plateau, black soil region in the northeast, and karst region in the southwest and regions along the "Belt and Road"; and steps to prevent sandstorm at source and curb stony desertification in Beijing and Tianjin. Desertificated land must be refrained from further development, and comprehensive steps must be taken to conserve water and soil in slope farmland and control gully erosion. Another effective way is to better manage the environment in small watersheds by controlling pollution and strengthening its ecological functions. One of the targets is to bring under control a total of 270,000 km2 of water and soil eroded land.

7. Protection and restoration of rivers, lakes and wetland

China will enhance the protection of natural wetlands in the upper and middle reaches of the Yangtze River, in the Yellow River and in Caohai Lake of Guizhou Province, better protect and manage those wetlands whose ecological functions were impaired and biodiversity is declining, and carry out demonstration projects on sustainable use of wetland. In addition, there will be steps to strengthen the protection of rare and endangered aquatic organisms, important aquatic genetic resources and important fishery waters such as spawning grounds, feeding grounds, wintering grounds and migration routes. Efforts will also go to protecting and restoring ecosystems in important rivers, lakes and wetlands ranging from the "six rivers and five lakes" in the Beijing-Tianjin-Hebei region, to the "four lakes" in Hubei Province, the upstream of the Qiantang River, Caohai Lake, Liangzi Lake, Fenhe River, Hutuo River and Hongjiannao Lake. An ecologically friendly approach would be applied to protecting and managing pollution of rivers and lakes in urban cities,

8. Rescue and protection of endangered wildlife species

Steps must be taken to protect the habitats of rare and endangered wildlife species such as giant panda, crested ibis, tiger, leopard, Asian elephant, orchid plants, cycads and wild rice, and to improve their living environment, build in situ conservation areas, rescue and breeding centers and gene banks, and carry out conservation breeding and reintroduction programs. Rescue efforts will focus on wildlife species with extremely small populations that can hardly survive or reproduce and on critically endangered wildlife species. There would be investigation and survey on germplasm of rare and endangered wildlife species, rescue efforts to collect and preserve those resources, and steps to establish a germplasm resources bank (pool).

9. Biodiversity conservation

It's planned to investigate and assess biodiversity in priority areas of biodiversity conservation, build 50 biodiversity observation stations and 800 sampling areas, and establish a biodiversity database, a platform for assessment and early warning of biodiversity and a platform for species inspection and identification. China will delineate the lines of national nature reserves and clearly identify the land-related rights. Over 60% of national nature reserves must be built in strict accordance with relevant standards. And there would be efforts to build and strengthen ecological corridors, relocate people that live in the core areas and the buffer zones of nature reserves step by step, refine the ex situ conservation system, and establish a national biodiversity museum. There would also be demonstration projects on biodiversity conservation and restoration as well as on poverty alleviation.

10. Prevention and control of invasive alien species (IAS)

China will select 50 typical national nature reserves for prevention and control of typical invasive alien species. It's planned to establish 50 IAS demonstration sites on prevention and control and utilization of IAS, 100 bases for breeding of IAS predators and 1000 km IAS isolation belt in provinces of Yunnan, Guangxi and southeastern coastal provinces that are severely plagued by alien species invasion. China will set up 300 port inspection stations, and make 50 major ports of entry more capable of preventing invasive alien species. As for the alien species already in China, there will be an IAS database in place based on detailed investigation, as well as an IAS monitoring and early warning system based on both satellite-based remote sensing and ground monitoring.

11. Improvement of forest quality

It's planned to expedite in cultivating mixed forests, nurturing forest plantations, restoring degraded forests, managing and protecting forests for public interest and breeding quality forest seeds. Quality of forests must be improved in major river sources, state-owned forest fields (farms) and collectively-owned forest fields. In addition, there would be 40 million hectare of forests well-tended and 9 million hectare of degraded forests restored.

12. Protection of heritage trees

China will strictly protect crown cover areas and root zones of heritage trees, with efforts to set up protection signs and fence. There would be stimulus measures to boost the growth of those withered and endangered tree species, in an effort to secure a total of 600,000 heritage trees and rejuvenate another 3 million ones.

13. Restoration of urban ecosystem and supply of ecological products

There will be investigation and evaluation of natural resources and ecological space in the planned urban area so as to identify areas that are damaged, with poor resilience, and in urgent need of restoration. Targeting the said areas, there would be demonstration projects restore ecosystems in urban areas. It's also planned to promote green roads and corridors; appropriately design and build green parks, upgrade old parks and increase the supply of ecological products.

14. Technological innovation in protecting ecological environment

China will develop a number of technological innovation platforms to protect ecological environment and prioritize the development of a number of specialized environmental high-tech development zones, and push forward major special research projects concerning water, air, soil, ecology, risk, and smart environmental protection. A range of innovation projects will be implemented, including those on improving regional environmental quality in the Beijing-Tianjin-Hebei region, in the Yangtze River Economic Belt, regions along the "Belt and Road Initiative", in the old industrial base of Northeast China, and in Xiangjiang River basin; those on protection and restoration of ecological-security shields in regions of Tibetan Plateau, Loess Plateau, shelterbelts in North China and karst region in the southwest; on safe disposal and recycling of urban waste; on prevention and control of environmental risks and clean alternatives; as well as on smart environment. Moreover, supporting systems must also be put in place including building key laboratories, engineering technology centers, scientific observatories and decision support system for environmental protection. In addition, China will set up the Lancang-Mekong River water resource cooperation center and environmental cooperation center as well as an information sharing and decision-making platform for the "Belt and Road Initiative".

Chapter 10 Introduce Supporting Measures to Facilitate Implementation

Section 1 Clarify responsibilities and tasks

Clarifying local objectives and responsibilities. Local governments are the main body responsible for implementing the *Plan*. They should integrate the objectives, tasks, measures and major projects on protection of ecological environment into regional economic and social development plan, and develop and make public the priority tasks and annual targets of environmental protection. Provinces, autonomous regions and municipalities must report to the public on its progress in implementing the *Plan*, and promote public participation and oversight, to ensure all tasks are completed.

Promoting coordinated efforts to finish planning tasks. While performing respective duties, the relevant authorities should closely cooperate in refining relevant systems and mechanisms, strengthening funding support, and thoroughly implementing the *Plan*. There must be cooperation and coordination mechanisms in place in fields of air, water, soil, heavy metals and biodiversity to discuss and address major issues in the aforementioned areas on a regular basis. The Ministry of Environmental Protection will report to the State Council on major progress in environmental protection annually.

Section 2 Increase inputs

Expanding financial input. Pursuant to the principle of matching powers of government authority with its expenditure responsibilities, and clearly defining the respective powers and responsibilities of the central and local governments, a financial management system for environmental protection must be put in place at an early date to meet the needs of environmental protection. It's encouraged to utilize special funds for environmental protection in an innovative way, for example, supporting third party treatment and public-private partnerships (PPPs) in pollution control. Funds for ecological conservation and restoration must be further integrated in response to addressing pollution and improving environment in mountains, waters, forests and farmlands in a systematic way.

Broadening financing channels. The user-pay system will be further refined to support business projects on environmental protection. In an effort to widely promote PPP model, China will actively advocate business projects on environmental protection, explore appropriate ways that employ the benefits of resource development projects and comprehensive resource utilization to compensate the input in pollution control projects and private investment so as to attract private investment in quasi-public welfare and public welfare environmental protection projects. China will encourage the establishment of market-oriented environmental defense fund by private capital. In addition, China will encourage more investment in environmental protection from venture capital companies, equity investment companies and social donations.

Section 3 Expand international cooperation

Participating in international environmental governance. China will actively engage in developing rules of global environmental governance, and in negotiating international environmental conventions, international nuclear safety convention as well as environment-related international trade and investment agreements; assume its international responsibilities as a large developing country and implement international conventions; regulate the activities of overseas environmental organizations that are operating in China in accordance with law; strengthen international communications and spread China's best practice and experience on environmental protection; and increase foreign aid in an innovative way based on the integrated plan on foreign aid.

Stepping up international cooperation. China will put in place and refine the mechanism of communication and cooperation with relevant countries, international organizations, research institutions and civil groups, and push forward dialogues and exchanges on advanced environmental protection visions, management systems & policies as well as environmental industries and technologies, in a view to bring China's environmental protection more aligned with the international level. China will launch a number of international cooperation projects on issues related to air, water, soil and biodiversity. It will implement the 2030 Agenda for Sustainable Development of the United Nations, and strengthen dialogue and pragmatic cooperation on environmental protection and nuclear safety with foreign countries, regions and international organizations. Moreover, China will enhance South-South cooperation, actively cooperate on ecological conservation, environmental protection and nuclear safety, and crack down on illegal trade on chemicals and illegal transboundary movement of solid waste.

Section 4 Push ahead pilot and demonstration projects

Developing national pilot zones for ecological conservation. Aiming at improving environmental quality and promoting green development, China will establish a number of national pilot zones for ecological conservation that would focus on institutional innovation, institutional supply and on exploring new models on it. China will actively promote "Green Cells" projects which call for building green communities, promoting sustainable development-based schooling, and establishing ecological industrial parks. It is hoped by 2017 major reforms in these pilot zones would be able to produce a number of implementable and effective systems to advance ecological civilization, and by 2020 a full-fledged institutional systems on ecological civilization would be established in the pilot zones, which can be replicated and promoted nationwide.

Demonstration projects playing a guiding role. China will develop demonstration zones on ecological civilization that are guided by a set of standards and operated under certain mechanisms. It's also required to take into account regional balance and China's environmental protection priorities in developing these zones. As they are in place, there would be follow-up supervision and management, as well as performance evaluation and experience review, to develop a replicable model that could be drawn upon, promoted and applied on a wider scope.

Promoting policy demonstration projects. China will carry out pilot projects on comprehensive reform and innovation for rural environmental protection system in which the attainment areas and non-attainment areas will be subject to pollution source supervision and discharge permit management according to emissions standards and environmental quality standards respectively. As part of the reform efforts, China will also implement pilot projects on environmental auditing, environmental damage compensation, environmental services and government procurement of services. For these projects, necessary policies would be introduced to facilitate the reform and strengthen supervision, and a third treatment of environmental pollution will cover a larger area and involve more sectors. China will also launch pilot projects on comprehensive reform of environmental protection at provincial level.

Section 5 Examine and evaluate progress and performance with strict standards

MEP in cooperation with relevant departments will conduct regular investigations on the performance of provinces, autonomous regions and municipalities in improving the environmental quality, controlling major pollutants discharge, and in implementing key projects on protecting ecological environment, and all of these investigation results will be made public. Based on all these performance review and evaluation, there will be a mid-term evaluation and final performance review on implementing the *Plan* respectively scheduled by then end of 2018 and 2020. The results will be reported to the State Council and made public, and taken as an important gauge when evaluating the performances of local leadership as well as relevant officials.